Photocatalysts for Organics Degradation

Photocatalysts for Organics Degradation

Special Issue Editors

Barbara Bonelli
Maela Manzoli
Francesca S. Freyria
Serena Esposito

MDPI • Basel • Beijing • Wuhan • Barcelona • Belgrade • Manchester • Tokyo • Cluj • Tianjin

Special Issue Editors

Barbara Bonelli
PoliTO BiomED
Interdepartmental Lab
Italy

Maela Manzoli
Università degli Studi di Torino
Italy

Francesca S. Freyria
Massachusetts Institute of
Technology
USA

Serena Esposito
University of Cassino and
Southern Latium
Italy

Editorial Office
MDPI
St. Alban-Anlage 66
4052 Basel, Switzerland

This is a reprint of articles from the Special Issue published online in the open access journal *Catalysts* (ISSN 2073-4344) (available at: https://www.mdpi.com/journal/catalysts/special_issues/organics_degradation).

For citation purposes, cite each article independently as indicated on the article page online and as indicated below:

LastName, A.A.; LastName, B.B.; LastName, C.C. Article Title. *Journal Name* **Year**, *Article Number*, Page Range.

ISBN 978-3-03928-286-9 (Pbk)
ISBN 978-3-03928-287-6 (PDF)

© 2020 by the authors. Articles in this book are Open Access and distributed under the Creative Commons Attribution (CC BY) license, which allows users to download, copy and build upon published articles, as long as the author and publisher are properly credited, which ensures maximum dissemination and a wider impact of our publications.

The book as a whole is distributed by MDPI under the terms and conditions of the Creative Commons license CC BY-NC-ND.

Contents

About the Special Issue Editors vii

Barbara Bonelli, Maela Manzoli, Francesca S. Freyria and Serena Esposito
Photocatalysts for Organics Degradation
Reprinted from: *Catalysts* **2019**, *9*, 870, doi:10.3390/catal9100870 1

Chunjing Hao, Zehua Xiao, Di Xu, Chengbo Zhang, Jian Qiu and Kefu Liu
Saturated Resin Ectopic Regeneration by Non-Thermal Dielectric Barrier Discharge Plasma
Reprinted from: *Catalysts* **2017**, *7*, 362, doi:10.3390/catal7120362 3

Zedong Zhu, Muthu Murugananthan, Jie Gu and Yanrong Zhang
Fabrication of a Z-Scheme $g-C_3N_4/Fe-TiO_2$ Photocatalytic Composite with Enhanced Photocatalytic Activity under Visible Light Irradiation
Reprinted from: *Catalysts* **2018**, *8*, 112, doi:10.3390/catal8030112 19

Honghui Pan, Wenjuan Liao, Na Sun, Muthu Murugananthan and Yanrong Zhang
Highly Efficient and Visible Light Responsive Heterojunction Composites as Dual Photoelectrodes for Photocatalytic Fuel Cell
Reprinted from: *Catalysts* **2018**, *8*, 30, doi:10.3390/catal8010030 35

Liang Jiang, Yizhou Li, Haiyan Yang, Yepeng Yang, Jun Liu, Zhiying Yan, Xiang Long, Jiao He and Jiaqiang Wang
Low-Temperature Sol-Gel Synthesis of Nitrogen-Doped Anatase/Brookite Biphasic Nanoparticles with High Surface Area and Visible-Light Performance
Reprinted from: *Catalysts* **2017**, *7*, 376, doi:10.3390/catal7120376 49

Vinh Huu Nguyen, Trinh Duy Nguyen, Long Giang Bach, Thai Hoang, Quynh Thi Phuong Bui, Lam Dai Tran, Chuong V. Nguyen, Dai-Viet N. Vo and Sy Trung Do
Effective Photocatalytic Activity of Mixed Ni/Fe-Base Metal-Organic Framework under a Compact Fluorescent Daylight Lamp
Reprinted from: *Catalysts* **2018**, *8*, 487, doi:10.3390/catal8110487 59

Byung-Geon Park
Photocatalytic Behavior of Strontium Aluminates Co-Doped with Europium and Dysprosium Synthesized by Hydrothermal Reaction in Degradation of Methylene Blue
Reprinted from: *Catalysts* **2018**, *8*, 227, doi:10.3390/catal8060227 79

About the Special Issue Editors

Barbara Bonelli (Professor of Chemistry Fundamentals for the Technologies). BB holds a PhD in Chemistry from the Università degli Studi di Torino (Italy) and has been enrolled at Politecnico di Torino (Italy) since April 2001. Her main scientific interests are the physico-chemical aspects related to heterogeneous catalysis and gas adsorption, the characterization of materials by means of spectroscopic techniques and other surface techniques. She is the co-author of more than 150 papers in peer-reviewed international journals (h-index 34).

Maela Manzoli Professor of Industrial Chemistry at the Department of Drug Science and Technology of the University of Turin, Italy). Her studies focus on the surface properties of polycrystalline solids of catalytic interest (DRUV-Vis, FTIR spectroscopy, nitrogen physisorption), as well as their textural, morphological and structural characterization (SEM, HRTEM, XRD, XANES, EXAFS) under reaction conditions. Particular interest is dedicated to supported noble metal nanoparticles, applied to a variety of catalytic processes assisted by MW, US or mechanochemistry. She is the co-author of three book chapters and about 120 papers in peer-reviewed international Journals (h-index 38).

Francesca S. Freyria, after an M.Eng. in Environmental Engineering, received the European Ph.D. degree in Materials Science and Technology at Politecnico of Torino (Italy) under the supervision of Professor B. Bonelli. In 2014, she joined Professor Bawendi's group at Massachusetts Institute of Technology (Cambridge, USA) as postdoc researcher with an MIT Energy Initiative fellowship. In 2019 she won a Marie Skłodowska-Curie Individual Fellowship to develop new hybrid antenna nanomaterials for artificial photosynthesis. Her broader research interests include the study of new heterostructured nanomaterials and mesoporous materials, and how to endow them with new properties for environmental remediation and solar energy applications.

Serena Esposito is an Assistant Professor at the Department of Applied Science and Technology, Politecnico di Torino (Italy). Her research activities deal with the definition of synthesis strategies to prepare nanomaterials with tailored physico-chemical features. Porous, magnetic, ceramic or metal-ceramic nanomaterials are mostly obtained by the sol-gel technique, and they are used in catalysis, fuel cells, biological separations and water remediation.

Editorial

Photocatalysts for Organics Degradation

Barbara Bonelli [1,*], Maela Manzoli [2], Francesca S. Freyria [1,3] and Serena Esposito [1]

1. Institute of Chemistry, Department of Applied Science and Technology; PoliTO BiomED Interdepartmental Lab, Politecnico di Torino, 10129 Torino, Italy; francesca.freyria@polito.it (F.S.F.); serena_esposito@polito.it (S.E.)
2. Department of Drug Science and Technology, Università degli Studi di Torino, Via Pietro Giuria 9, 10125 Torino, Italy; maela.manzoli@unito.it
3. Department of Chemistry, Massachusetts Institute of Technology, Cambridge, MA 02139, USA
* Correspondence: barbara.bonelli@polito.it

Received: 11 October 2019; Accepted: 14 October 2019; Published: 21 October 2019

Organics degradation is one of the challenges of Advanced Oxidation Processes (AOPs), which are mainly employed for the removal of water and air pollutants. Compared to stand-alone processes, AOPs are more effective if combined with other removal means, especially to degrade recalcitrant (stable) pollutants in subsequent steps.

Integrated systems able to solve the aforementioned issues in a single step could be less expensive and more efficient, but their development requires advancements from the point of view of both materials and the process. In this Issue, a system consisting of integrated resin adsorption/Dielectric Barrier Discharge (DBD) plasma regeneration was proposed to treat some textile dyes, showing that the DBD plasma could maintain the efficient adsorption performance of the resin while degrading the adsorbed dye [1].

Some AOPs imply the presence of catalyst, especially in photocatalytic processes: the goal of photocatalysis is to find efficient and stable materials (under the reaction conditions), which are able both to stabilize excitons (i.e., the generated electron/hole pairs) and to exploit solar light. However, the last two goals remain very ambitious and require breakthrough advances from the point of materials science (synthesis methods) and physical chemistry. Moreover, a multi-technique approach could help in studying the surface and textural properties of the photocatalyst in order to be able to unravel the phenomena regulating excitons formation and stabilization.

Different solutions are reported in the literature, including the production of nanocomposites [2,3] and of mixed phases of TiO_2 [4]. The former have to be properly designed, like in Z-Scheme $g-C_3N_4$/Fe-TiO_2 [2] for the photodegradation of phenol, and in heterojunction nanostructured composites for photocatalytic fuel cells [3]: both systems were able to absorb in the Vis region. As a whole, the formation of heterojunctions in the nanocomposites simultaneously favors the photogenerated electron/hole separation and maintains the reduction and oxidation properties.

Occurrence of mixed phases is another means to promote and stabilize excitons, like in Degussa P25, where the mixed rutile/anatase phase is considered responsible for its good photocatalytic performance. Recently, mixed TiO_2 phases containing brookite have been proved to display improved photocatalytic efficiency, like in N-doped anatase/brookite nanoparticles [4], obtained with high surface area by a low temperature sol-gel technique. Again, the development of new nanomaterials has been shown to have an impact on the progress of photocatalytic efficiency. Such advancements may be obtained by a plethora of synthesis methods, leading to different nanomaterials, like mixed Ni/Fe-based Metal Organic Frameworks (MOFs) [5] and Sr aluminates co-doped with Eu and Dy [6]. The former are porous networks, with high specific surface areas, where a thorough physico-chemical characterization by multiple techniques showed [5] that the occurrence of mixed-metal cluster Fe_2NiO was able to enhance the photocatalytic performance further, via an electron transfer effect. The latter materials were instead prepared by different methods, namely with a hydrothermal reaction at low T and using a sol-gel method [6], pointing out the importance of developing new synthetic routes to obtain engineered (nano)materials for photocatalytic applications.

References

1. Hao, C.; Xiao, Z.; Xu, D.; Zhang, C.; Qiu, J.; Liu, K. Saturated Resin Ectopic Regeneration by Non-Thermal Dielectric Barrier Discharge Plasma. *Catalysts* **2017**, *7*, 362. [CrossRef]
2. Zhu, Z.; Murugananthan, M.; Gu, J.; Zhang, Y. Fabrication of a Z-Scheme g-C_3N_4/Fe-TiO_2 Photocatalytic Composite with Enhanced Photocatalytic Activity under Visible Light Irradiation. *Catalysts* **2018**, *8*, 112. [CrossRef]
3. Pan, H.; Liao, W.; Sun, N.; Murugananthan, M.; Zhang, Y. Highly Efficient and Visible Light Responsive Heterojunction Composites as Dual Photoelectrodes for Photocatalytic Fuel Cell. *Catalysts* **2018**, *8*, 30. [CrossRef]
4. Jiang, L.; Li, Y.; Yang, H.; Yang, Y.; Liu, J.; Yan, Z.; Long, X.; He, J.; Wang, J. Low-Temperature Sol-Gel Synthesis of Nitrogen-Doped Anatase/Brookite Biphasic Nanoparticles with High Surface Area and Visible-Light Performance. *Catalysts* **2017**, *7*, 376. [CrossRef]
5. Nguyen, V.H.; Nguyen, T.D.; Bach, L.G.; Hoang, T.; Bui, Q.T.P.; Tran, L.D.; Nguyen, C.V.; Vo, D.N.; Do, S.T. Effective Photocatalytic Activity of Mixed Ni/Fe-Base Metal-Organic Framework under a Compact Fluorescent Daylight Lamp. *Catalysts* **2018**, *8*, 487. [CrossRef]
6. Park, B. Photocatalytic Behavior of Strontium Aluminates Co-Doped with Europium and Dysprosium Synthesized by Hydrothermal Reaction in Degradation of Methylene Blue. *Catalysts* **2018**, *8*, 227. [CrossRef]

© 2019 by the authors. Licensee MDPI, Basel, Switzerland. This article is an open access article distributed under the terms and conditions of the Creative Commons Attribution (CC BY) license (http://creativecommons.org/licenses/by/4.0/).

Article

Saturated Resin Ectopic Regeneration by Non-Thermal Dielectric Barrier Discharge Plasma

Chunjing Hao, Zehua Xiao, Di Xu, Chengbo Zhang, Jian Qiu and Kefu Liu *

Department of Light Sources & Illuminating Engineering, Fudan University, Shanghai 200433, China; 17110720032@fudan.edu.cn (C.H.); zhxiao14@fudan.edu.cn (Z.X.); 15210720020@fudan.edu.cn (D.X.); 13110720012@fudan.edu.cn (C.Z.); jqiu@fudan.edu.cn (J.Q.)
* Correspondence: kfliu@fudan.edu.cn; Tel.: +86-21-5566-5184

Received: 26 October 2017; Accepted: 22 November 2017; Published: 27 November 2017

Abstract: Textile dyes are some of the most refractory organic compounds in the environment due to their complex and various structure. An integrated resin adsorption/Dielectric Barrier Discharge (DBD) plasma regeneration was proposed to treat the indigo carmine solution. It is the first time to report ectopic regeneration of the saturated resins by non-thermal Dielectric Barrier Discharge. The adsorption/desorption efficiency, surface functional groups, structural properties, regeneration efficiency, and the intermediate products between gas and liquid phase before and after treatment were investigated. The results showed that DBD plasma could maintain the efficient adsorption performance of resins while degrading the indigo carmine adsorbed by resins. The degradation rate of indigo carmine reached 88% and the regeneration efficiency (RE) can be maintained above 85% after multi-successive regeneration cycles. The indigo carmine contaminants were decomposed by a variety of reactive radicals leading to fracture of exocyclic C=C bond, which could cause decoloration of dye solution. Based on above results, a possible degradation pathway for the indigo carmine by resin adsorption/DBD plasma treatment was proposed.

Keywords: indigo carmine; resin; Dielectric Barrier Discharge; adsorption; regeneration

1. Introduction

Industrial production processes, especially in printing and dyeing manufacturing, generate large quantities of refractory wastewater every year [1–3]. Organic chemical dyestuffs are applied as coloring material in textile industry, and are hard to degrade in normal ways, such as adsorption [4,5], biological [6], and chemical methods [7,8]. These methods have many disadvantages. On the one hand, in biological treatment, is difficult to cultivate suitable active species. On the other hand, chemical disposal often brings the problem of secondary pollution. In addition, the physical adsorption method is only a phase transfer of contaminants, and adsorbents are usually by chemical regeneration, resulting in secondary contamination of chemical reagents. Hence, systems of advanced oxidation processes (AOPs) with conventional approaches are integrated for the decomposition of organic dye contaminants, such as Fenton oxidation process (H_2O_2 + Fe^{2+}) [9], ozone and UV (O_3 + UV), photocatalytic oxidation (TiO_2 + UV), and Non-thermal plasma (NTP) [10–12]. These methods are mainly based on the strong oxidative properties of hydroxyl radical and degradation of organic molecules. The history of Dielectric Barrier Discharge can be traced back to 1857 [13]. In 1987, Sidney first presented the technique of high voltage pulsed discharge to dispose of sewage [14]. After that, many research teams have studied the applications in various areas [15–17]. DBD plasma is widely used in the environmental protection because it produces a large number of high energy electrons, intense UV radiation, and a variety of chemical free radicals (e.g., hydroxyl radical, high energy oxygen atoms, etc.), which can rapidly react with most of the bio-refractory organic pollutants. Nevertheless, DBD plasma technology alone needs higher energy consumption, and wastewater quality factors, such as concentration, conductivity, and pH value, greatly affect the degradation effect. In particular, relatively low concentration and Chemical Oxygen Demand (COD) may greatly waste discharge energy, and more energy probably heats the wastewater solution. Hence, one of the most promising technologies, combining adsorption and DBD plasma to degrade pollutants, was introduced. Research on highly concentrated pollutants and regeneration for saturated adsorbent has been reported [18,19]. At present, the adsorbents applied in the wastewater treatment are Granular Activated Carbon (GAC), zeolite, activated alumina, etc. [20,21]. However, although Activated Carbon (AC) has been widely used in the industry, the adsorption performance of saturated AC greatly decreases after multiple regeneration. Moreover, the regeneration of AC is difficult, for example the use of heating regeneration method resulting in high carbon loss rate, or the use of pharmaceutical regeneration method resulting in high costs and secondary pollution. AC is also conductive, which is not conducive to DBD plasma discharge [22]. Based on some related literature [23–25], resin has strong adsorption properties, and can keep strong ability to absorb contaminants through repeated regeneration. The general regenerative method is eluted by mixed solution, which can lead to chemical secondary pollution [26]. Unlike the general methods, DBD discharge regeneration can achieve double effect, in which concentrated pollutants onto resins are decomposed and saturated resins are regenerated to restore the adsorption performance. At present, no literature has mentioned the study on regeneration of resin by plasma. In this article, we conducted an in-depth study that confirmed the combination of resin adsorption and DBD regeneration process can greatly improve the degradation efficiency of pollutants and reduce operating costs.

In this paper, a flat-plate reactor to investigate a facile wastewater treatment technique was designed. There are five aspects researched: (1) the adsorption behavior of resin about indigo carmine solution; (2) the regeneration efficiency for multiple cycles; (3) the functional groups and structure properties of resin before and after DBD plasma treatment; (4) the analysis of intermediate products in gas and liquid phase; and (5) a possible degradation pathway of indigo carmine contaminants by resin adsorption/DBD plasma discharge treatment system. The technique of integrated resin adsorption/DBD plasma regeneration method has very broad prospects in the field of environment protection.

2. Results

Degradation Pathway Process

The possible reaction pathway for the degradation of indigo carmine solution by absorption/DBD plasma regeneration system was proposed (Figure 1). The pathway included all of the detected intermediates and showed the active radicals as oxidant, especially the hydroxyl radical formed in the DBD discharge process. Other weak oxidants were also possible, such as H_2O_2 and HO_2. According to the LC-MS analysis results, isatin 5-sulfonic acid (m/z 226) was the main aromatic product produced when a hydrogen radical attacked the C=C bond of indigo carmine. Isatin 5-sulfonic acid then lost SO_4^{2-} and converted to isatin. Further oxidation of the intermediate products led to a mixture of carboxylic acid and amine. Finally, those carboxylic acid and amine were degraded to inorganic molecule, including of carbon dioxide, ammonium, nitrate, etc.

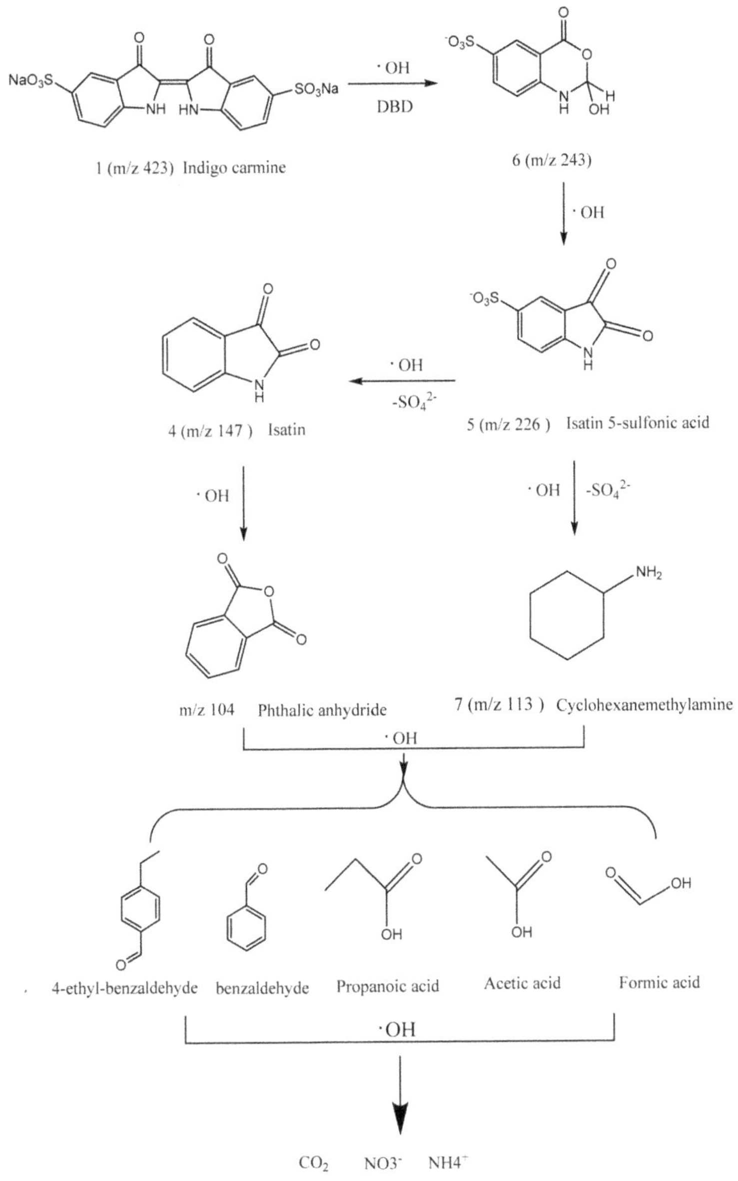

Figure 1. Degradation pathway of indigo carmine in an integrated resin adsorption/DBD plasma.

3. Discussion

3.1. Effect of Regeneration on Adsorption Capacity and Kinetics of Resin

By comparing the adsorption isotherms of virgin resin with a series of DBD regenerated resins, the effect of DBD plasma on the adsorption capacity was analyzed. Figure 2 depicts the adsorption isotherms of indigo carmine on virgin and series of adsorption/DBD regenerated resins. It was observed that the adsorption capacity of regenerated resin is reduced, and, as the regeneration cycle progresses, the q_e value of the resin samples decreased.

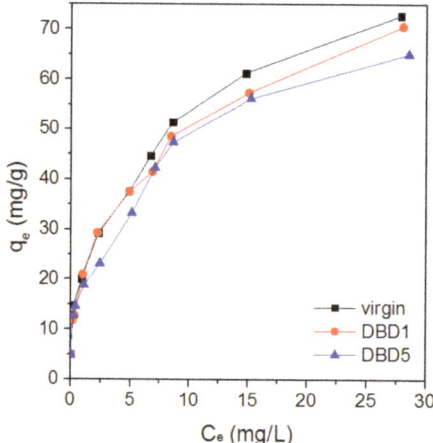

Figure 2. Adsorption isotherms of indigo carmine on virgin and different saturated/DBD regeneration resins.

On the other hand, the adsorption type of indigo carmine onto resin samples after DBD plasma treatment was also studied. Generally, the Freundlich model was a kind of adsorption isotherm model, which was generally expressed by Freundlich equation (see, e.g., [27]):

$$q_e = K_F C_e^{1/n}, \qquad (1)$$

where q_e is the amount of adsorption equilibrium state, mg/g; C_e is the concentration of equilibrium solution, mg/L; K_F (L/g) is the Freundlich parameter interaction with adsorption and adsorption capacity; and the exponential term of $1/n$ (dimensionless) is related to the adsorption force. ln q_e and ln C_e plotted in a straight line from the slope and intercept of the straight line were the values of $1/n$ and ln K_F, respectively. The fitting curve of the linear correlation coefficient was R^2. The above three constants are listed in Table 1. The results showed that all isotherms fitted well to the Freundlich equation, which indicated that regeneration process did not seem to alter adsorption processes. All $1/n$ values were less than 1, which indicated further adsorption of indigo carmine onto resins. The adsorption isotherms of indigo carmine onto resins confirmed this phenomenon (Figure 2).

Table 1. Freundlich constants for adsorption of indigo carmine onto resin.

Sample	K_F (L/g)	$1/n$	R^2
Virgin	2.31	0.33	0.989
DBD1	2.39	0.30	0.979
DBD5	2.29	0.30	0.989

3.2. Effect of Regeneration on the Regeneration Efficiency

The residual concentration changes of indigo carmine were analyzed on virgin and DBD regenerated resins (Figure 3). Five DBD treated cycle experiments were conducted. The first to fifth DBD plasma regeneration experiments were abbreviated as DBD1, DBD2, DBD3, DBD4, and DBD5, respectively. There was only a little change in adsorption rate for DBD regenerated resins, demonstrating that adsorption rate almost kept the same level after five cycles of regeneration. Hence, the DBD regeneration efficiency could directly reveal the impact of DBD discharge process, which was calculated using the following Equation (2):

$$RE = \frac{q_r}{q_v} \times 100\%, \qquad (2)$$

where q_v and q_r are the amounts of adsorption equilibrium state of indigo carmine on virgin and regenerated resins, respectively (mg/g).

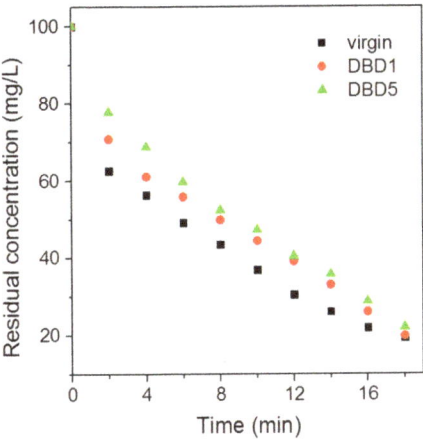

Figure 3. Residual concentration changes of indigo carmine solutions by virgin and different saturated/DBD regeneration resins.

All regeneration efficiencies of this process by series of regeneration cycles are presented in Figure 4. The residual concentration of the indigo carmine solution adsorbed by the virgin and regenerated resin was basically achieved, which was less than 20%. At the same time, it was also observed that, as the number of regeneration cycles increased, the degradation efficiency remained almost unchanged, indicating that the structural properties of resins remained stable and DBD plasma did not cause serious damage to the active sites on the surface of resins (discussed below). As can be seen in Figure 4, the regeneration efficiency of the resin was maintained at 80% or more even after five regeneration cycles. The experimental setup was high voltage value of 18 kV, current of 4.32 A, frequency of 1 kHz, and the degradation rate of 86%. The energy efficiency of resin adsorption/DBD plasma treatment was 139.5 g/kWh, whereas the DBD plasma treating the same concentration of indigo carmine was 56.5 g/kWh, based on the previous work [2]. The energy efficiency of adsorption/DBD regeneration was greater than 2.5 times the DBD plasma system.

The UV-Vis spectra of the resin samples at each treatment cycle are shown in Figure 5. The peaks were caused by the residual indigo carmine and intermediates onto resins after DBD regeneration. The wavelengths of 610, 450, 280, and 250 nm onto regenerated resins were observed before and after treatment of the UV-Vis spectra. The wavelength of 610 nm was characteristic absorption peak of indigo carmine. Moreover, the chromophoric group and unsaturated bond of indigo carmine correspond to the wavelengths of 610 and 250 nm, respectively. The formula of the indigo carmine is shown in the inset of Figure 5, and the bond in the bracket is the chromophoric group. The absorption intensity of all of the peaks decreased through every regeneration cycles. These results showed that chromophoric and unsaturated bonds of indigo carmine were almost broken up, which illuminated that the saturated resin was regenerated sufficiently, maintaining great degradation efficiency after multiple successive discharge cycles, which is a promising technique.

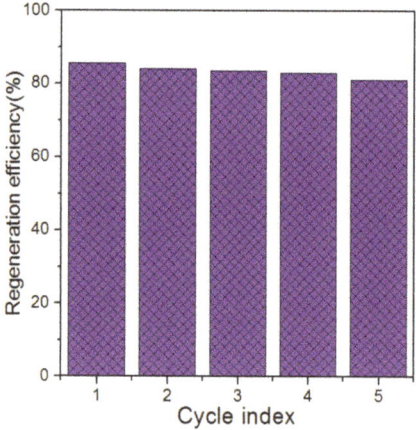

Figure 4. The regeneration efficiencies of resins after DBD plasma multiple cycles.

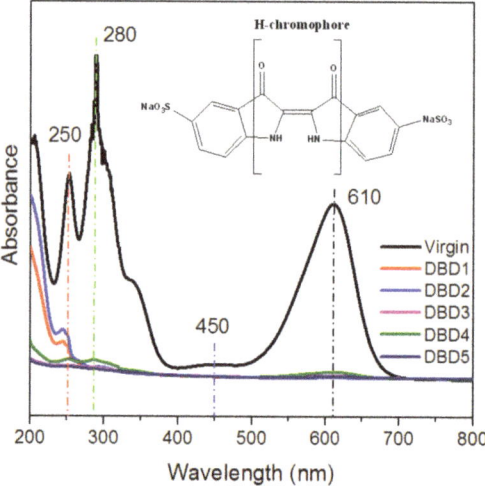

Figure 5. The UV-Vis spectra of virgin and DBD regeneration treatment resins.

3.3. Changes in the Structural Properties of Resins

The chemical bonds of the resin were characterized with FT-IR (fourier transform infrared spectroscopy) spectrometer. The FT-IR spectra of the three kinds of resins sample, containing virgin, saturated adsorbed resin, and adsorbed/DBD plasma regenerated resin, are depicted in Figure 6. The peaks at the wavelength of ~3420, ~2940, ~1650, ~1450, ~1100, and ~680 cm^{-1} for all of the resin samples indicated that the resin surface functional groups were not destroyed. The broadening bond around the main peak, ~3420 cm^{-1}, could be mainly caused by O-H stretching vibration peak in water [25]. The peak at ~3420 cm^{-1} was a multi-absorption peak, which was widened by overlapping with nitrogen hydrogen bond (N-H) and O-H stretching vibration peaks. The absorption peak at ~2940 cm^{-1} was mainly caused by porogen (polyethylene glycol), and residual organic liquid paraffin on the resin surface. The N-H bending vibration absorption peak corresponded to the position at ~1650 cm^{-1}. The band of 1450 cm^{-1} was primarily linked to the aromatic ring of C=C functional groups. At the peak of 1100 cm^{-1}, it was generally matched with C-O stretching in the lactate and ether groups [28]. The adsorption intensity of the saturated adsorbed resins had enhanced compared with the virgin resin, which demonstrated clearly that adsorbed contaminants onto resins could increase the intense of hydrogen bond, double bond of carbon, and carbon oxygen bond. After the multiple regenerative cycles plasma treatment, the intensity of all the absorption peaks were decelerated compared with saturated resin, which was possible on account of the adsorbed indigo carmine onto resin achieved a certain degree of degradation. Note that the bond around 680 cm^{-1} attributed C-N bonds of indigo carmine were cleaved partially and indigo carmine was decomposed to some decolorized intermediates [29].

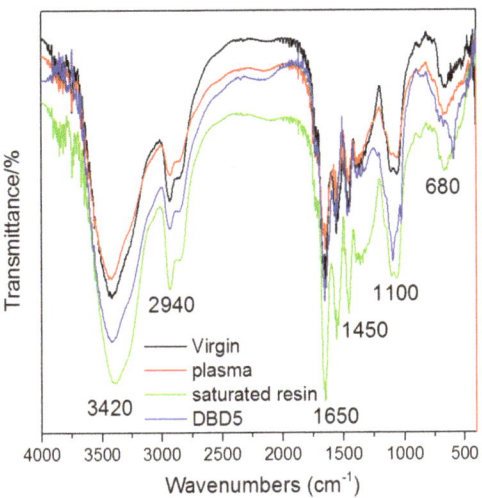

Figure 6. FT-IR spectra of the four kinds of virgin, plasma, saturated and DBD5 resins.

Apart from the analysis of functional groups, the structural characteristics of virgin, saturated, and DBD5 resins are listed in Table 2. The analysis showed that virgin and DBD5 resins exhibited similar specific surface area, total pore volume, pore size, and adsorption capacity. The analysis showed that DBD plasma regeneration process did not destroy the structure of resin. Therefore, the reason of the reduced performance of the saturated resin was that the adsorbed organic molecules occupied the adsorption site.

Table 2. Structural characteristics of virgin, saturated, and regenerated samples.

Sample	S_{BET} (m^2/g)	$V_{Total\ pore}$ (cm^3/g)	Pore Size (nm)	Adsorption Capacity (mg/g)
Virgin	163	0.3210	43.05	-
Saturated	74.6	0.1630	23.94	70.58
DBD5	157.8	0.299	46.33	65.06

3.4. Identification of Intermediates by GC-MS and LC-MS

As shown in Figure 7, the GC-MS (gas chromatography-mass spectrometer) analysis exhibited six peaks related to formic acid (m/z = 29) at t_r = 1.6 min, acetic acid (m/z = 43) at t_r = 1.75 min, benzaldehyde (m/z = 105) at t_r = 3.17 min, octamethyl-cyclotetrasiloxane (m/z = 281) at t_r = 4.5 min, 4-ethyl-benzaldehyde (m/z = 134) at t_r = 5.63 min, and phthalic anhydride (m/z = 104) at t_r = 6.45 min. The peak position was almost identical to a previous study [30]. Note that carboxylic acids came from the heterocyclic ring opening of isatin-5-sulfonic acid sodium salt dihydrate, which was been confirmed during DBD plasma degradation of indigo carmine. Aldehyde and acid anhydride could be formed from the oxidation of their CO-NH-CO groups. Octamethyl was a kind of siloxane copolymer, which was probably formed when silica wool was heated. The cooperation of the plasma with the resin would still produce some intermediate products. The results are listed in Table 3.

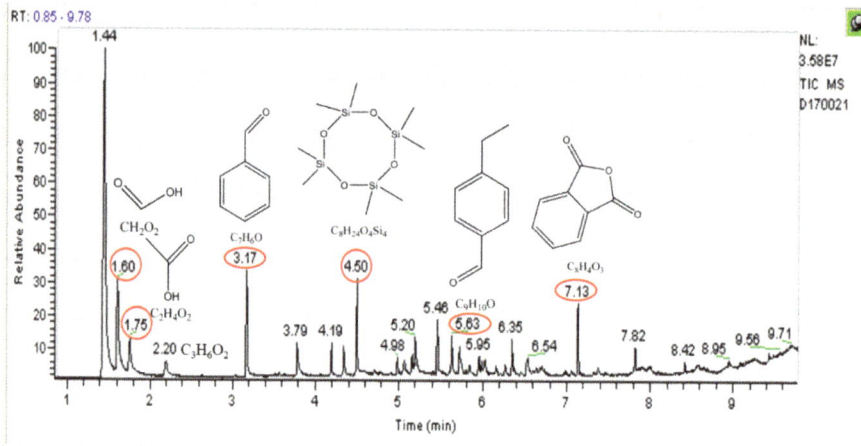

Figure 7. Total ion chromatogram of decomposed compositions by GC–MS analysis.

Table 3. Analysis of degradation products by GC-MS.

Compound	Structure	Molecular Formula	Retention Time (min)
Formic acid		CH_2O_2	1.6
Acetic acid		$C_2H_4O_2$	1.75
Propanoic acid		$C_3H_6O_2$	2.2
Benzaldehyde		C_7H_6O	3.17
Octamethyl-cyclotetrasiloxane		$C_8H_{24}O_4Si_4$	4.5
4-ethyl-benzaldehyde		$C_9H_{10}O$	5.63
Phthalic anhydride		$C_8H_4O_3$	7.13

Figure 8 displays the LC-MS (liquid chromatography-mass spectrometer) analysis of the indigo carmine solution and the molecular formula of indigo carmine. Note that main charged anion of m/z

423 and its isotopic variants including m/z 423.9 (m+1) and 425 (m+2) well fitted those calculated for $C_{16}H_8N_2O_8S_2$. The anion of m/z 423 was detected as the primary species in dying solution.

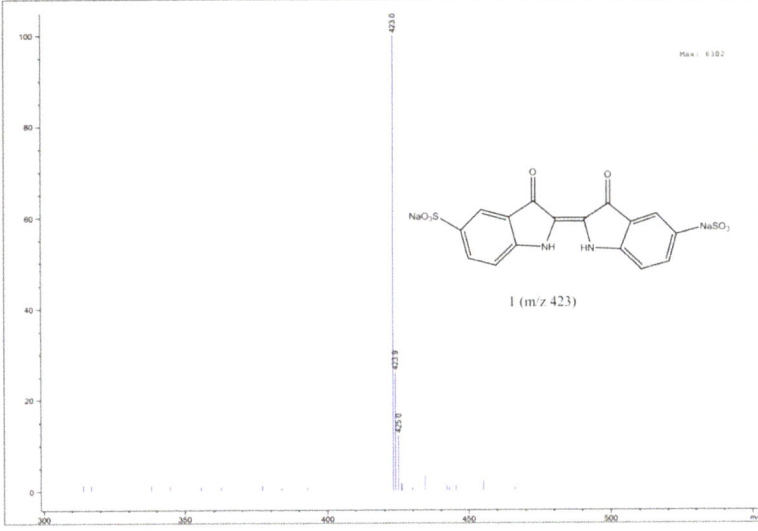

Figure 8. LC-MS analysis of initial indigo carmine solution.

The LC-MS analysis of the indigo carmine aqueous solution adsorbed by virgin resin and the molecular formula of the predominant component is shown in Figure 9. Whereas the anion of m/z 423 was not detected, ions of m/z 228.2, 229.3, 250.3, and 338.5 were clearly observed. Obviously, the components of m/z 228.2 (m+1) and m/z 229.3 (m+2) were isotopologs of isatin 5-sulfonic acid with molecular formula of $C_8H_5NO_5S$ [31]. The cation of m/z 250.3 was an isotopolog of 5-Isatinsulfonic acid sodium salt, which proved the fracture of C=C bond. Based on these results, the continuous formation of intermediates adsorbed onto virgin resin had a much smaller π-electron conjugated system than the initial molecule, which could result in the indigo carmine solution decoloration, as experimentally observed. To analyze the residual pollutants on the surface of the plasma regenerated resin, LC-MS of indigo carmine solution adsorbed onto resin regenerated by DBD plasma and the molecular formula of the main byproducts are shown in Figure 10. The anion of m/z 226.1 is doubly charged, as evidenced by the presence of the (M+1) isotopologs of m/z 226. The $\triangle m/z$ for the doubly charged anions m/z 226.1 and 243.9 was 18 units, which indicated the latter molecule could be formed from the former via the incorporation of two hydroxyl groups. The anion of m/z 113.1 could probably be fitted with cyclohexylmethanamine with molecular formula of $C_7H_{15}N$. The peaks at other locations might be caused by residual surfactant on the surface of resin. Therefore, the formation of these intermediate products in aqueous solution was owing to fracture of the chromophoric C=C group and incorporation of oxygen atoms, hydroxyl groups, etc. Hence, the analysis of indigo carmine degraded by DBD plasma treatment by LC-MS allowed us to detect unknown byproducts and analyze the degradation pathway in the reaction. Soem of the intermediates in aqueous solution by resin adsorption/DBD plasma regeneration are listed in Table 4.

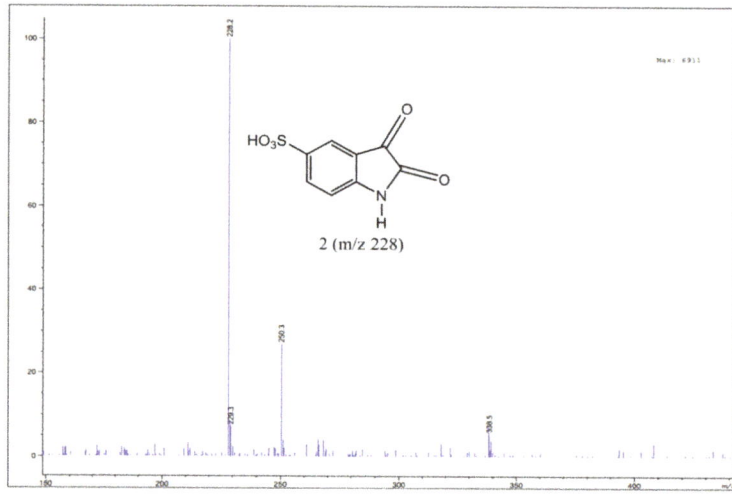

Figure 9. LC-MS analysis of indigo carmine solution adsorbed by virgin resin sample.

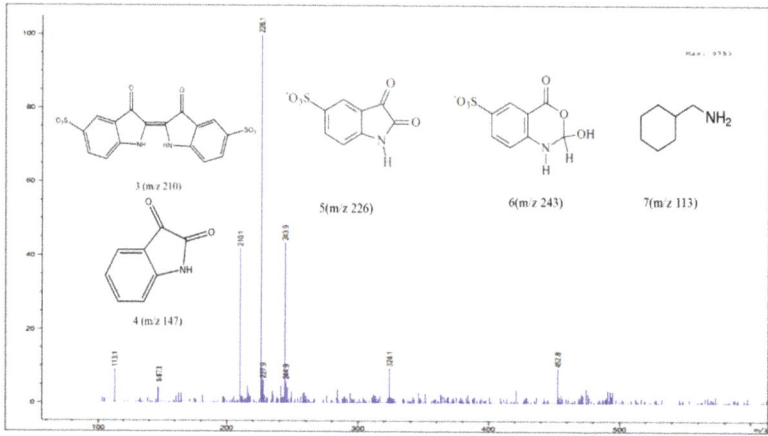

Figure 10. LC-MS analysis of indigo carmine solution adsorbed by plasma regenerated sample.

Table 4. Analysis of degradation products by LC-MS.

Compound	Structure	Molecular Formula	MS Fragments (m/z)
Indigo carmine		$C_{16}H_8N_2Na_2O_8S_2$	423/425
		$C_{16}H_8N_2O_8S_2$	210
5-Isatinsulfonic acid sodium salt		$C_8H_4NO_5S \cdot Na$	250

Table 4. Cont.

Compound	Structure	Molecular Formula	MS Fragments (m/z)
Isatin 5-sulfonic acid		$C_8H_5NO_5S$	228/229
		$C_8H_4NO_5S$	226/227
		$C_8H_6NO_6S$	243/244
Isatin		$C_8H_5NO_2$	147
Cyclohexanemethylamine		$C_7H_{15}N$	113

4. Materials and Analysis Methods

4.1. Materials

The resins used in this experiment were manufactured by Shaanxi LanShen Special Resin Factory, China. The type of the resin was LS-109D. The resins were pretreated based on the following steps: Firstly, the resin was soaked in the anhydrous ethanol for 24 h and washed with ethanol mixed with water in a volume ratio of 1:5 until the effluent was clear with the absence of ethanol. Secondly, the above resin was soaked in 4% hydrochloric acid for 2 h and washed to neutral with deionized water, and then in 4% sodium hydrate soaked for 2 h and washed to neutral with deionized water. Finally, the resin was dried at 60 °C to constant weight and placed in the dryer for reserve. The initial resin is abbreviated as Virgin. Indigo carmine was purchased from the Sinojpharm Chemical Reagent Co., Ltd. (Shanghai China). The analytical grade of all other reagents was used in the experiment (Aladdin Reagent Co., Ltd. Shanghai China). The concentration of 1000 mg/L stock solutions was made by dissolving indigo carmine powder into deionized water. The adopted concentrations in the adsorption experiment were acquired by diluting the stock solution with deionized water.

4.2. The DBD Regeneration Reaction System

The schematic diagram of DBD regeneration system is shown in Figure 11. It primarily included pulsed power supply, oxygen cylinder, and the regeneration reactor. The schematic diagram of the adsorbent-packed DBD reactor is shown in Figure 12. It was a flat type of DBD reactor. The ground electrode and high voltage electrode of DBD reactor was copper sheet. The discharge electrode was placed onto quartz barrier (80 mm × 30 mm × 2 mm). The discharge gap space between the ground and high voltage electrode was kept at 3 mm. In addition, the flat reactor port filled with quartz wool was to prevent the resin from blowing out by the carrier gas during the discharge process. As the carrier gas of oxygen, the flow rate was 3 L/min. The discharge voltage and current waveforms were recorded with the oscilloscope (Tektronix TDS 2014, Johnston, OH, USA), with a voltage probe (Tektronix P6021, Johnston, OH, USA) and a current probe (Tektronix P6021, Johnston, OH, USA), which are shown in Figure 13. The discharge parameters in the regeneration process were pulse frequency of 1 kHz, pulse voltage of 16 kV, current of ~4.3 A, storage capacitance C_p of 3.8 pF, and reaction time of 5 min, with a high rise time of about 50 ns.

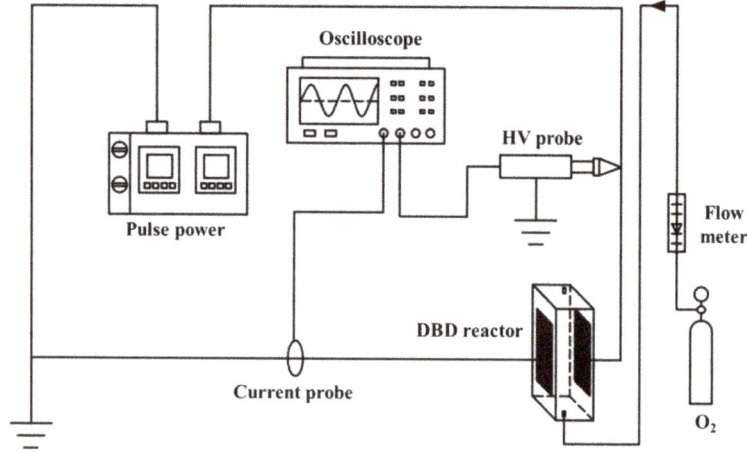

Figure 11. Bench-scale apparatus of DBD plasma for Indigo Carmine decomposition.

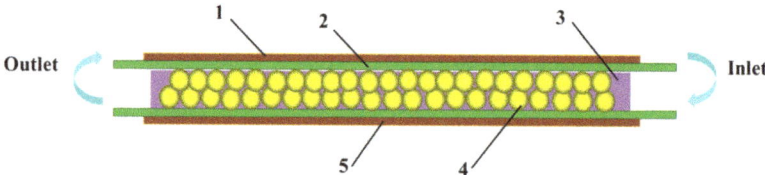

Figure 12. Schematic diagram of the DBD regeneration reactor: 1, high voltage electrode; 2, quartz glass; 3, plasma area; 4, resin; and 5, ground electrode.

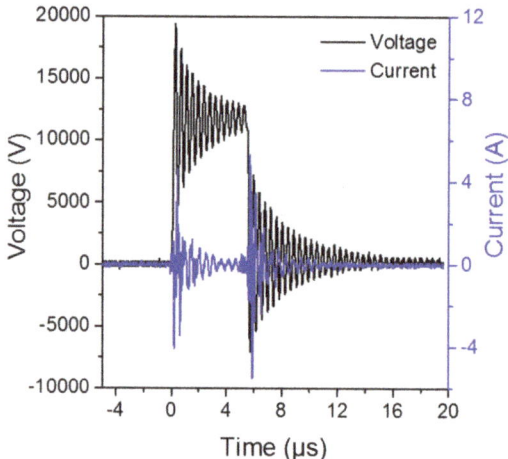

Figure 13. Voltage and current waveforms delivered to the DBD regeneration reactor.

4.3. Analytical Device

The pH of solutions was measured with a FE20 meter (Mettler Toledo, Greifensee, Switzerland). The concentration of indigo carmine was detected by UV-Vis spectrophotometer (Specord® 200 Plus, Analytikjena, Jena, Germany) using the supernatant from the filtered solution and detection at the maximum wavelength of 610 nm. The analytical samples separated from treated solution were filtered with Whatman 0.45 µm PTFE membrane filter before analysis. The intermediates of samples were analyzed by Liquid Chromatograph Mass Spectrometer (Agilent 6400, Agilent Technologies Inc., Santa Clara, CA, USA) analysis with a C18 column and ultraviolet detection at 610 nm. The mobile phase was the volume ratio of 7:3 (v/v) between acetonitrile and deionized water (with 0.01% formic acid) with a flow rate of 1 mL/min. Furthermore, the filtrate of samples was extracted thrice with dichloromethane and evaporated in a vacuum evaporator (BUCHI R-300, Buchi, Flawil, Switzerland) with 40 °C water bath, after which a gas chromatography (Agilent 6890N) coupled with a mass selective (Agilent 5975) apparatus and a capillary column (30 mm × 0.25 mm × 0.25 mm) was utilized for identification of byproducts in gas phase during the regeneration process. The functional groups of virgin, saturated and DBD regenerated resin were a by Fourier transform-infrared (FT-IR) spectroscopy. The analytic samples were prepared by mixing 1 mg of the samples with 500 mg of KBr in an agate mortar and scanned in a range from 4000 to 400 cm^{-1}. The structural properties of resin were obtained from the physical adsorption of N2 at 350 K determined by a Tristar II 3020 equipment. The special surface area was calculated using the BET equation [32]. To evaluate the adsorption capacity, the adsorption equilibrium isotherms of indigo carmine onto resins were measured based on the method provided by Mangun [33].

4.4. The Regeneration of Indigo Carmine Saturated Resin

The regeneration reaction of resins was carried out in the DBD reactor. Before the regeneration process, 0.25 g of saturated resins were put into the reactor. The regeneration reaction started when the power supply was open, which would last for 10 min. The resins were regenerated for five cycles in total. The first to fifth DBD plasma regeneration experiments were abbreviated as DBD1, DBD2, DBD3, DBD4, and DBD5, respectively. During DBD plasma regeneration process, the reaction temperature was not more than 35 °C. All experiments were carried out at atmospheric pressure.

4.5. Kinetics Adsorption

The kinetics adsorption reaction of indigo carmine onto virgin and regenerated resins were operated in oscillatory reactor. The concentration of solution after adsorption was monitored by the mentioned method.

4.6. Adsorption Equilibrium Isotherms

The adsorption isotherms of indigo carmine onto virgin and regenerated resins were operated in oscillatory reactor. Exactly 0.25 g of resins were added into a series of conical bottles containing 100 mL of indigo carmine solution of different concentration. The concentration was 5, 10, 15, 20, 30, 40, 50, 60, 80, and 100 mg/L, respectively. The conical flasks with cover were shaken with a constant speed of 120 rpm at 40 °C for 12 h. Then, the suspension was filtered for further analysis. Based on the standard curves of indigo carmine samples, the concentration was analyzed with UV–Vis spectrophotometer, and the amount of indigo carmine adsorbed onto resins was inferred from Equation (3):

$$q_e = \frac{(C_0 - C_e)V}{m}, \qquad (3)$$

where q_e is the amount of indigo carmine adsorbed per gram of resin, mg/g; V is the volume of the liquid phase, L; C_0 is the concentration of the initial solution before it contacts with resin, mg/L; C_e is the concentration of the solution at equilibrium condition, mg/L; and m is the amount of the resin, g.

5. Conclusions

An integrated system of resin adsorption/DBD plasma regeneration method was applied for the degradation of indigo carmine solution. According to the GC-MS and LC-MS analytical results, above 85% of indigo carmine adsorbed on resin was decomposed into sulfonic acid and dehydroxylation byproducts by DBD plasma. Simultaneously, saturated resin was regenerated, and the adsorption capacity of adsorption/DBD plasma regenerated resin could be maintained at a relatively high level after multiple cycles. The functional groups, specific surface area, total pore volume, pore size, and adsorption capacity of regenerated resin did not suffer a large degree of damage. The multiple cycles of regenerative reaction indicated that resin maintained a stable and effective performance for indigo carmine adsorption. Finally, the possible degradation pathway of indigo carmine was proposed in the resin adsorption/DBD plasma regeneration process. This integrated method has a good prospect in the treatment of refractory organic wastewater.

Acknowledgments: The work was supported by Chinese National Nature Science Foundation under Grant 11075041.

Author Contributions: Chunjing Hao conceived, designed and performed the experiments; Zehua Xiao processed the reactor; Di Xu, Chengbo Zhang, Jian Qiu contributed reagents/materials/analysis tools, and Kefu Liu mainly summarized and refined the analysis; and Chunjing Hao wrote the paper.

Conflicts of Interest: The authors declare no conflict of interest.

References

1. Vandevivere, P.C.; Bianchi, R.; Verstraete, W. Treatment and reuse of wastewater from the textile wet-processing industry: Review of emerging technologies. *J. Chem. Technol. Biotechnol.* **1998**, *72*, 289–302. [CrossRef]
2. Minamitani, Y.; Shoji, S.; Ohba, Y.; Higashiyama, Y. Decomposition of dye in water solution by pulsed power discharge in a water droplet spray. *IEEE Trans. Plasma Sci.* **2008**, *36*, 2586–2591. [CrossRef]
3. Jiang, S.; Wen, Y.; Liu, K. Treatments of dye wastewater by water spout in the pulsed DBD. In Proceedings of the IEEE Internatyional Conference on Plasma Science (ICOPS), Antalya, Turkey, 24–28 May 2015.
4. Khraisheh, M.A.M.; Al-Ghouti, M.A.; Allen, S.J.; Ahmad, M.N. Effect of OH and silanol groups in the removal of dyes from aqueous solution using diatomite. *Water Res.* **2005**, *39*, 922–932. [CrossRef] [PubMed]
5. Wu, X.; Hui, K.N.; Hui, K.S.; Lee, S.K.; Zhou, W.; Chen, R.; Hwang, D.H.; Cho, Y.R.; Son, Y.G. Adsorption of basic yellow 87 from aqueous solution onto two different mesoporous adsorbents. *Chem. Eng. J.* **2012**, *180*, 91–98. [CrossRef]
6. Crini, G. Non-conventional low-cost adsorbents for dye removal: A review. *Bioresour. Technol.* **2006**, *97*, 1061–1085. [CrossRef] [PubMed]
7. Li, X.Z.; Zhang, M. Decolorization and biodegradability of dyeing wastewater treated by a TiO_2-sensitized photo-oxidation process. *Water Sci. Technol.* **1996**, *34*, 49–55.
8. Azbar, N.; Yonar, T.; Kestioglu, K. Comparison of various advanced oxidation processes and chemical treatment methods for COD and color removal from a polyester and acetate fiber dyeing effluent. *Chemosphere* **2004**, *55*, 35–43. [CrossRef] [PubMed]
9. Raghu, S.; Basha, C.A. Chemical or electrochemical techniques, followed by ion exchange, for recycle of textile dye wastewater. *J. Hazard. Mater.* **2007**, *149*, 324–330. [CrossRef] [PubMed]
10. Garcia-Montano, J.; Ruiz, N.; Munoz, I.; Domenech, X.; Garcia-Hortal, J.A.; Torrades, F.; Peral, J. Environmental assessment of different photo-Fenton approaches for commercial reactive dye removal. *J. Hazard. Mater.* **2006**, *138*, 218–225. [CrossRef] [PubMed]
11. Fongsatitkul, P.; Elefsiniotis, P.; Yamasmit, A.; Yamasmit, N. Use of sequencing batch reactors and Fenton's reagent to treat a wastewater from a textile industry. *Biochem. Eng. J.* **2004**, *21*, 213–220. [CrossRef]
12. Pouran, S.R.; Aziz, A.R.A.; Daud, W.M.A.W. Review on the main advances in photo-Fenton oxidation system for recalcitrant wastewaters. *J. Ind. Eng. Chem.* **2015**, *21*, 53–69. [CrossRef]
13. Kogelschatz, U. Dielectric-barrier discharges: Their history, discharge physics, and industrial applications. *Plasma Chem. Plasma Process.* **2003**, *23*, 1–46. [CrossRef]

14. Clements, J.S.; Sato, M.; Davis, R.H. Preliminary investigation of pre-breakdown phenomena and chemical-reactions using a pulsed high voltage discharge in water. *IEEE Trans. Ind. Appl.* **1987**, *23*, 224–235. [CrossRef]
15. Eliasson, B.; Kogelschatz, U. Modeling and applications of silent discharge plasmas. *IEEE Trans. Ind. Appl.* **1991**, *19*, 309–323. [CrossRef]
16. Laroussi, M. Nonthermal decontamination of biological media by atmospheric-pressure plasmas: Review, analysis, and prospects. *IEEE Trans. Plasma Sci.* **2002**, *30*, 1409–1415. [CrossRef]
17. Vergara Sanchez, J.; Torres Segundo, C.; Montiel Palacios, E.; Gomez Diaz, A.; Reyes Romero, P.G.; Martinez Valencia, H. Degradation of textile dye AB 52 in an aqueous solution by applying a plasma at atmospheric pressure. *IEEE Trans. Plasma Sci.* **2017**, *45*, 479–484. [CrossRef]
18. Kuroki, T.; Fujioka, T.; Kawabata, R.; Okubo, M.; Yamamoto, T. Regeneration of honeycomb zeolite by nonthermal plasma desorption of toluene. *IEEE Trans. Ind. Appl.* **2009**, *45*, 10–15. [CrossRef]
19. Jiang, B.; Zheng, J.; Lu, X.; Liu, Q.; Wu, M.; Yan, Z.; Qiu, S.; Xue, Q.; Wei, Z.; Xiao, H.; et al. Degradation of organic dye by pulsed discharge non-thermal plasma technology assisted with modified activated carbon fibers. *Chem. Eng. J.* **2013**, *215*, 969–978. [CrossRef]
20. Hao, X.L.; Zhang, X.W.; Lei, L.C. Degradation characteristics of toxic contaminant with modified activated carbons in aqueous pulsed discharge plasma process. *Carbon* **2009**, *47*, 153–161. [CrossRef]
21. Berenguer, R.; Marco-Lozar, J.P.; Quijada, C.; Cazorla-Amoros, D.; Morallon, E. Electrochemical regeneration and porosity recovery of phenol-saturated granular activated carbon in an alkaline medium. *Carbon* **2010**, *48*, 2734–2745. [CrossRef]
22. Dabrowski, A.; Podkoscielny, P.; Hubicki, Z.; Barczak, M. Adsorption of phenolic compounds by activated carbon—A critical review. *Chemosphere* **2005**, *58*, 1049–1070. [CrossRef] [PubMed]
23. Lin, S.; Juang, R. Adsorption of phenol and its derivatives from water using synthetic resins and low-cost natural adsorbents: A review. *J. Environ. Manag.* **2009**, *90*, 1336–1349. [CrossRef] [PubMed]
24. Iyim, T.B.; Acar, I.; Oezguemues, S. Removal of basic dyes from aqueous solutions with sulfonated phenol-formaldehyde resin. *J. Appl. Polym. Sci.* **2008**, *109*, 2774–2780. [CrossRef]
25. Fan, J.; Li, H.; Shuang, C.; Li, W.; Li, A. Dissolved organic matter removal using magnetic anion exchange resin treatment on biological effluent of textile dyeing wastewater. *J. Environ. Sci.* **2014**, *26*, 1567–1574. [CrossRef] [PubMed]
26. Wang, C.; Lippincott, L.; Meng, X. Feasibility and kinetics study on the direct bio-regeneration of perchlorate laden anion-exchange resin. *Water Res.* **2008**, *42*, 4619–4628. [CrossRef] [PubMed]
27. Foo, K.Y.; Hameed, B.H. Insights into the modeling of adsorption isotherm systems. *Chem. Eng. J.* **2010**, *156*, 2–10. [CrossRef]
28. Sundaraganesan, N.; Anand, B.; Meganathan, C.; Joshua, B.D. FT-IR, FT-Raman spectra and ab initio HF, DFT vibrational analysis of 2,3-difluoro phenol. *Spectrochim. Acta A* **2007**, *68*, 561–566. [CrossRef] [PubMed]
29. Ahmed, M.A.; Brick, A.A.; Mohamed, A.A. An efficient adsorption of indigo carmine dye from aqueous solution on mesoporous Mg/Fe layered double hydroxide nanoparticles prepared by controlled sol-gel route. *Chemosphere* **2017**, *174*, 280–288. [CrossRef] [PubMed]
30. Reddy, P.M.K.; Raju, B.R.; Karuppiah, J.; Reddy, E.L.; Subrahmanyam, C. Degradation and mineralization of methylene blue by dielectric barrier discharge non-thermal plasma reactor. *Chem. Eng. J.* **2013**, *217*, 41–47. [CrossRef]
31. Wang, J.; Lu, L.; Feng, F. Improving the indigo carmine decolorization ability of a bacillus amyloliquefaciens laccase by site-directed mutagenesis. *Catalysts* **2017**, *7*, 275. [CrossRef]
32. Qu, G.-Z.; Li, J.; Wu, Y.; Li, G.-F.; Li, D. Regeneration of acid orange 7-exhausted granular activated carbon with dielectric barrier discharge plasma. *Chem. Eng. J.* **2009**, *146*, 168–173. [CrossRef]
33. Mangun, C.L.; Benak, K.R.; Economy, J.; Foster, K.L. Surface chemistry, pore sizes and adsorption properties of activated carbon fibers and precursors treated with ammonia. *Carbon* **2001**, *39*, 1809–1820. [CrossRef]

© 2017 by the authors. Licensee MDPI, Basel, Switzerland. This article is an open access article distributed under the terms and conditions of the Creative Commons Attribution (CC BY) license (http://creativecommons.org/licenses/by/4.0/).

Article

Fabrication of a Z-Scheme g-C$_3$N$_4$/Fe-TiO$_2$ Photocatalytic Composite with Enhanced Photocatalytic Activity under Visible Light Irradiation

Zedong Zhu [1], Muthu Murugananthan [2], Jie Gu [1] and Yanrong Zhang [1,*]

1. Environmental Science Research Institute, Huazhong University of Science and Technology, Wuhan 430074, China; zedong_zhu@hust.edu.cn (Z.Z.); wang_zhao@hust.edu.cn (J.G.)
2. Department of Chemistry, PSG College of Technology, Peelamedu, Coimbatore 641004, India; muruga.chem@gmail.com
* Correspondence: yanrong_zhang@hust.edu.cn; Tel.: +86-027-87792107-802

Received: 30 January 2018; Accepted: 8 March 2018; Published: 13 March 2018

Abstract: In the present study, a nanocomposite material g-C$_3$N$_4$/Fe-TiO$_2$ has been prepared successfully by a simple one-step hydrothermal process and its structural properties were thoroughly studied by various characterization techniques, such as X-ray diffraction (XRD), Fourier Transform Infrared (FTIR) spectroscopy, electron paramagnetic resonance (EPR) spectrum, X-ray photoelectron spectroscopy (XPS), and UV-vis diffuse reflectance spectrometry (UV-vis DRS). The performance of the fabricated composite material towards the removal of phenol from aqueous phase was systematically evaluated by a photocatalytic approach and found to be highly dependent on the content of Fe^{3+}. The optimum concentration of Fe^{3+} doping that showed a dramatic enhancement in the photocatalytic activity of the composite under visible light irradiation was observed to be 0.05% by weight. The separation mechanism of photogenerated electrons and holes of the g-C$_3$N$_4$/Fe-TiO$_2$ photocatalysts was established by a photoluminescence technique in which the reactive species generated during the photocatalytic treatment process was quantified. The enhanced photocatalytic performance observed for g-C$_3$N$_4$-Fe/TiO$_2$ was ascribed to a cumulative impact of both g-C$_3$N$_4$ and Fe that extended its spectrum-absorptive nature into the visible region. The heterojunction formation in the fabricated photocatalysts not only facilitated the separation of the photogenerated charge carriers but also retained its strong oxidation and reduction ability.

Keywords: titanium dioxide; graphitic carbon nitride; Fe doping; Z-scheme

---

1. Introduction

For the past several decades, many semiconducting materials have been employed as photocatalysts and their photocatalytic performance was proved to be appropriate for organic pollutant degradation, hydrogen production from water splitting, and the reduction of CO$_2$ into fuels [1–3]. Among the studied materials, titanium dioxide (TiO$_2$) has been widely investigated owing to its excellent photocatalytic performance, viability, nontoxic nature, and good chemical stability. However, TiO$_2$ badly suffers from its wide band gap (3.0–3.2 eV) and low quantum efficiency, which limits its efficiency in practical applications. The conventional drawback of TiO$_2$ as a photocatalyst is that it can be activated only in the ultraviolet light region. Hence, work on extending its absorptive behavior into the visible range and reducing the photoexcited electron–hole pair recombination rate has been carried out by several strategies, such as the doping of metal (Fe, Cu, V) [4–6] and nonmetal (N, S) [7,8] elements into the lattice of TiO$_2$, the deposition of a noble metal (Pt, Au) [9,10] on its surface as a cocatalyst, and coupling it with an another semiconductor to form a heterojunction structure that narrows the band gap of TiO$_2$.

The doping of transition metal into a semiconductor (TiO_2) is one of the effective approaches to extend its absorptive behavior into the visible range besides improving the quantum efficiency. In particular, Fe has been considered to be a suitable candidate as the radius of both Fe^{3+} and Ti^{4+} (Fe^{3+}: 0.69 Å; Ti^{4+}: 0.745 Å) is almost the same, so that the incorporation of Fe into the crystal lattice of TiO_2 becomes easier [11]. In addition, as the energy level of Fe^{2+}/Fe^{3+} is much closer to that of Ti^{3+}/Ti^{4+}, Fe^{3+} could provide a shallow trap for photo-generated charge carriers that favors charge separation, which in turn improves the quantum yield efficiency [11].

In an another approach to enhance the efficiency of charge separation in TiO_2, a heterojunction structure consisting of two different semiconductors has been demonstrated [12]. Once the heterojunction has formed between TiO_2 and the coupled semiconductor material of a suitable band gap, the photoexcited electron of the lower conduction band (CB) potential of TiO_2 will be promoted to the CB potential of the coupled semiconductor material, and similarly, the photoexcited hole of the higher valence band (VB) potential of TiO_2 will be transferred to the VB of the coupled semiconductor material [13]. The oxidation and reduction abilities of the composite comes from those of the transferred respective photoexcited carriers, which are weaker than those of the original counterparts. As a result, though the charge separation efficiency of the composite is improved by the heterojunction, the oxidation and reduction abilities of the composite are decreased considerably [14–16]. Nevertheless, a coupling of two different semiconductors could lead to a formation of a typical Z-scheme system, in which the photoexcited electrons from the semiconductor with a less negative CB will transfer to the VB of the coupled semiconductor and combine with the photoexcited holes over there [17,18]. A composite following the Z-scheme system exhibits a higher redox capability than either of the components alone, thereby enhancing the charge separation efficiency and increasing the lifetime of charge carries as well. Owing to the above-mentioned advantages, the work on developing TiO_2-based Z-scheme photocatalysts has emerged as an important research area in the recent past.

Ever since the debut work carried out on graphite carbon nitride (g-C_3N_4) in 2009 [19], the metal-free semiconductor has attracted the attention of scientists working in the photocatalytic domain due to its narrow band gap (2.7 eV), extreme negative CB position (−1.12 eV versus Normal Hydrogen Electrode (NHE)), structural flexibility, and good chemical stability. Although the activation of pristine g-C_3N_4 can be achieved in the visible light region up to 460 nm, its photocatalytic efficiency is limited due to the high recombination probability of photoexcited electron–hole pairs [20]. It is expected that coupling g-C_3N_4 with TiO_2 would form a Z-scheme photocatalytic system and solve the problems normally encountered when using each of the semiconducting materials individually. In order to further improve the photocatalytic performance of the g-C_3N_4/TiO_2 composite under visible-light irradiation, attempts on developing composites, such as g-C_3N_4-Ti^{3+}/TiO_2 and S-, N-, or Fe^{3+}-doped TiO_2/g-C_3N_4, have been made [21–24]. However, very few works have been done on the fabrication of g-C_3N_4 and Fe-doped TiO_2 nanocomposite structures. Phenol is one of the most common organic water pollutants, because it is toxic even at low concentrations, and also its presence in natural waters can lead further to the formation of substituted compounds during disinfection and oxidation processes. Additionally, phenol is a model non-dye pollutant and a typical refractory aromatic compound considered to be a good probe molecule in testing photocatalytic activity for environmental purposes. The photocatalytic abatement of phenol vapors on anatase TiO_2 and g-C_3N_4-Ti^{3+}/TiO_2 nanotubes has been the object of a study [25,26], and the mineralization process is complete in about 3–4 h and 7–8 h, respectively.

The present work focused on preparing a photocatalytic nanocomposite g-C_3N_4/Fe-TiO_2 by a simple one-step hydrothermal process followed by a complete characterization using instrumental techniques such as X-ray Diffraction (XRD), X-ray Photoelectron Spectroscopy (XPS), Scanning Electron Microscopy (SEM), Transmission Electron Microscopy (TEM), Electron Paramagnetic Resonance (ESR), and UV-vis Diffused Reflectance spectrophotometry (UV-vis DRS). The photocatalytic activity of the as-prepared g-C_3N_4/Fe-TiO_2 composites was investigated under the visible light region by preparative degradation experiments using phenol as model pollutant. To further confirm the enhanced activity

2. Results and Discussion

2.1. Phase Structures and Morphology

Figure 1 shows the XRD patterns of the TiO_2, Fe-TiO_2, g-C_3N_4/TiO_2 (CT), 0.05Fe-CT, and g-C_3N_4 samples. It can be seen that all of the TiO_2-based samples exhibit identical diffraction patterns. The 2θ peaks observed at 25.3°, 37.8°, 48.0°, 54.0°, and 62.4° were well-matched with the standard data and correspond to the (101), (004), (200), (204), and (211) crystal planes of anatase TiO_2, respectively [27]. The two prominent diffraction peaks observed at 13.6° and 27.7° for pure g-C_3N_4 could be attributed to the diffraction patterns of the (100) and (002) crystal planes, respectively [28]. No peak corresponding to the characteristics of g-C_3N_4 was observed in either the CT or 0.05Fe-CT samples, and this might be due to the relatively poor crystallization and less content of g-C_3N_4 within the composites [29]. In addition, there is no obvious change in the peaks of anatase TiO_2 in both the composites, which indicates that neither the coupling of g-C_3N_4 nor the Fe-doping affects the phase structure of TiO_2.

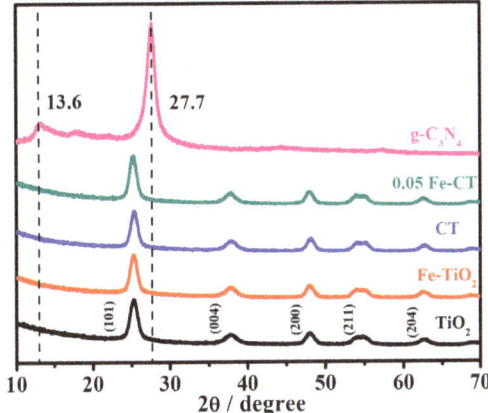

Figure 1. XRD patterns of TiO_2, Fe-TiO_2, CT, 0.05Fe-CT, and g-C_3N_4.

The images of SEM and TEM taken for the synthesized samples are displayed in Figures 2 and 3, respectively. The pure g-C_3N_4 sample exhibits a layered structure with a smooth surface that can be clearly seen in Figure 2a. This layered structure is expected to provide more sites for the growth of Fe-TiO_2 nanoparticles. Figure 2b reveals that Fe-TiO_2 materials consist of the aggregation of small nanocrystals. From the result shown in Figure 2c, it is found that the layered structure remains intact upon the incorporation of Fe-TiO_2 nanoparticles; moreover, the surface of g-C_3N_4 becomes slightly rough due to the formation of Fe-TiO_2 nanoparticles, suggesting that at least two semiconductors are in absolute physical contact with each other, which is the premise for the probable formation of either heterojunction or Z-scheme composites. Additionally, Energy Dispersive X-ray Spectroscopy (EDX) mapping of the composite shown in Figure 2d–h confirms the presence of Ti, O, C, N, and Fe elements in the 0.05Fe-CT sample. These results along with those of XRD and SEM suggest that Fe-TiO_2 nanoparticles are successfully loaded on the surface of g-C_3N_4.

Figure 2. SEM images (**a**) g-C$_3$N$_4$; (**b**) Fe-TiO$_2$; (**c**) 0.05Fe-CT; (**d–h**) EDX mapping of 0.05Fe-CT.

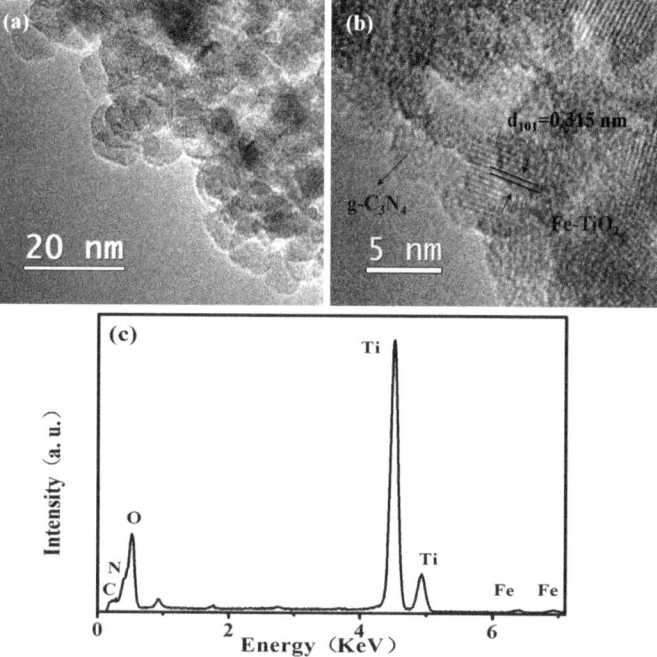

Figure 3. TEM (**a**,**b**) and EDX; (**c**) images of 0.05Fe-CT.

From the high resolution transmission electron microscopy (HR-TEM) image of the 0.05Fe-CT sample as shown in Figure 3a,b, a distribution of TiO$_2$ nanoparticles with the size of ~5 nm on the surface of g-C$_3$N$_4$ was confirmed. The lattice spacing of TiO$_2$ nanoparticles was found to be 0.351 nm, which matches with the (101) plane. The corresponding EDX pattern shows the existence of C, N, Ti, O, and Fe elements, which was in accordance with the results of the EDX mapping discussed earlier.

As seen in Figure 4a, the Fourier Transform Infrared (FTIR) spectrum for the g-C$_3$N$_4$, TiO$_2$, CT, and 0.05Fe-CT samples shows strong bands in the region of 450–4000 cm^{-1}. For g-C$_3$N$_4$, the bands observed around 1100–1650 cm^{-1} could be assigned to C-N and C=N stretching vibrations; the band at 810 cm^{-1} corresponds to s-triazinering vibrations; and the band around 3000–3300 cm^{-1} is correlated to

N-H stretching vibration modes [30–33]. For the TiO$_2$ sample, the band observed around 500–700 cm^{-1} could be accounted for with Ti-O stretching and Ti-O-Ti bridging stretching modes; and the bands at 1630 and 3400 cm^{-1} are correlated to the H-O-H bending stretch of surface-adsorbed water and its hydroxyl groups, respectively [34]. The characteristic bands observed for g-C$_3$N$_4$ and TiO$_2$ appeared in both the CT and 0.05Fe-CT composite samples too.

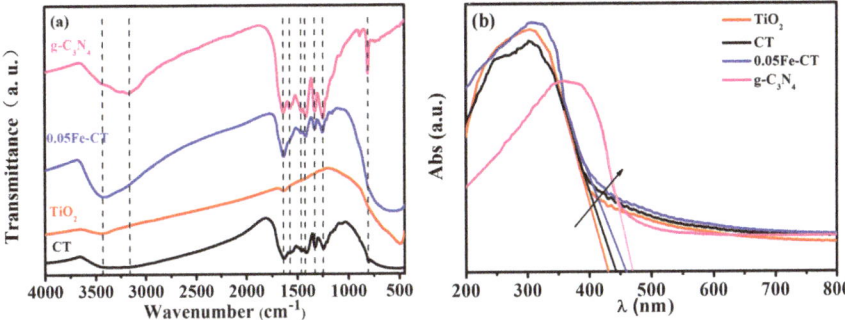

Figure 4. FTIR spectra (a) and UV-vis DRS; (b) of g-C$_3$N$_4$, TiO$_2$, CT, and 0.05Fe-CT.

As seen in Figure 4b, the optical property of the 0.05Fe-CT, CT, pure TiO$_2$, and g-C$_3$N$_4$ samples was measured by UV-vis diffuse reflectance spectroscopy. The UV-vis DRS spectra of 0.05Fe-CT and CT are quite similar to that of pure TiO$_2$, except for a slight movement of their main absorption edges toward the visible light region. In addition, upon the incorporation of g-C$_3$N$_4$, i.e., for the CT sample, a red shift moving up to 443 nm was observed, which indicates a reduced bandgap absorption edge of 2.80 eV estimated from 1240/λ (λ describes wavelength) [35]. Further, upon the doping of Fe to the CT sample, the light absorption of 0.05Fe-CT was extended to a still longer wavelength region and the appearance of the highest red shift, to a maximum of 461 nm (2.69 eV), was observed. These observations clearly indicate that the incorporation of composite material and the doping of Fe could graft a photocatalyst with the ability to utilize visible light effectively.

EPR as a highly sensitive spectroscopic technique for examining paramagnetic species can give valuable information about the lattice site wherein a paramagnetic doping ion is located. This technique can detect Fe ions to an extent of even less than 0.01 wt % in metal-oxide matrices [36]. The EPR spectra of TiO$_2$, CT, 0.01Fe-CT, 0.03Fe-CT, 0.05Fe-CT, and 0.06Fe-CT are depicted in Figure 5. At high magnetic field, a symmetrical EPR signal is observed at g = 2.004 for both TiO$_2$ and CT as well, which is an identification of the trapping of electrons on oxygen vacancies [37]. In addition, the EPR signal of CT is in accordance with that of TiO$_2$, which strongly indicates that the presence of CNs has no influence on the phase structure of the TiO$_2$. For the xFe-CT samples, unsymmetrical signals are observed at g = 1.99, which can be assigned to the fact that the Fe^{3+} is substituted for Ti^{4+} in the octahedral surroundings/atmosphere [37,38], otherwise it could simply be an overlapping of the two kinds of EPR signals. Further, as no signal at other g values was observed, the existence of Fe ions as Fe$_2$O$_3$-type clusters (g = 2.16) could not be possible [39,40]. It is worth noting that the intensity of signals at g = 1.99 of xFe-CT samples as shown in Figure 5a increases with increasing Fe^{3+} content, which indicates that the substitution of Fe^{3+} for Ti^{4+} in the TiO$_2$ lattice was effectively accomplished by a hydrothermal approach of simply increasing the iron content in the solution mixture. As seen in Figure 5b, the EPR spectra at low magnetic field exhibited very weak signals of g value at 4.29, which suggests that Fe^{3+} was located in a strongly distorted rhombic environment [40]. It is clear that the specific signals of EPR spectra at both high and low magnetic field confirmed the successful incorporation of Fe^{3+} into the crystal lattice of TiO$_2$ by a one-step hydrothermal method.

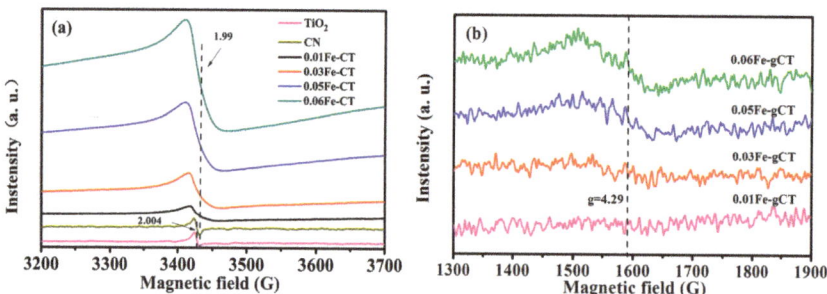

Figure 5. EPR spectra of different test field (a,b) of TiO$_2$, 0.01Fe-CT, 0.03Fe-CT, 0.05Fe-CT, and 0.06Fe-CT.

In order to examine the chemical states of elements involved in the as-prepared samples, XPS measurements were performed. The comparison of the Ti 2p spectra for samples TiO$_2$, Fe-TiO$_2$, and 0.05Fe-CT is shown in Figure 6a. The Ti 2p$_{3/2}$ and Ti 2p$_{1/2}$ peak positions of the TiO$_2$ sample were 458.55 eV and 464.25 eV, whereas they shifted to a higher binding energy of 458.65 eV and 467.35 eV for the Fe-TiO$_2$ and 0.05Fe-CT samples, respectively. The small shifts of binding energy might be due to the effect of the Fe^{3+} in the interstitial and/or substitutional site in the TiO$_2$ crystal lattice and formed the Ti-O-Fe bonds in the crystal lattice [41,42]. Due to the low doping level, the signals of Fe were too weak to be observed (not shown). As for the O 1s spectra presented in Figure 6b, two peaks of the binding energy at 529.95 and 532 eV for the 0.05Fe-CT sample were associated with the O$_2^-$ in TiO$_2$ and the -OH terminal on the surface [42]. For the 0.05Fe-CT sample, the formation of the new Ti-O-Fe bonds in the crystal lattice might change the electron densities of Ti^{4+} cations and O$_2^-$ anions, which caused a slightly higher shift of O 1s peaks compared to those for TiO$_2$ at 529.75 and 531.65 eV, respectively, and which might be a cause for the enhanced photocatalytic activity [42,43].

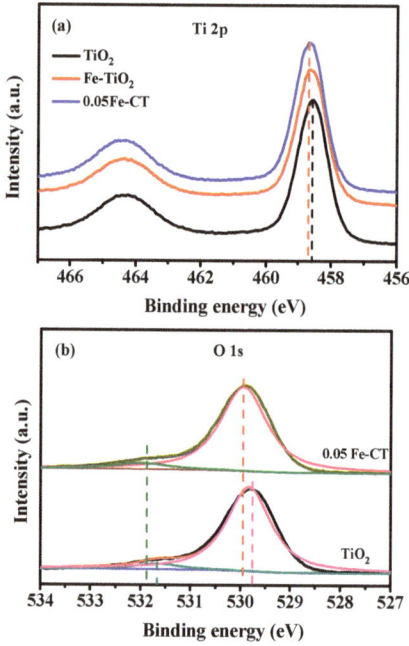

Figure 6. *Cont.*

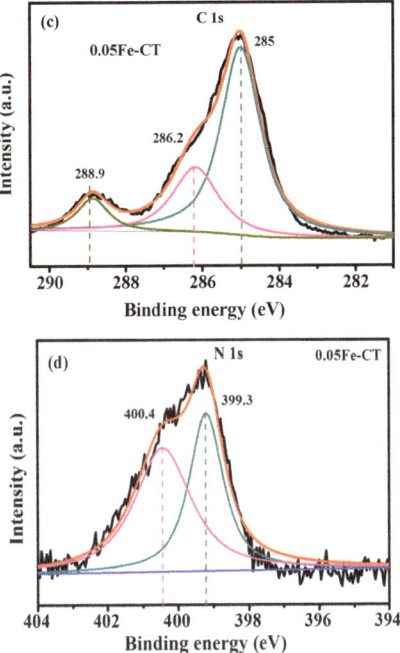

Figure 6. High resolution XPS spectra of (**a**) Ti 2p, (**b**) O 1s, (**c**) C 1s, and (**d**) N 1s.

From the XPS spectra of C 1s in Figure 6c, three peaks centered at 285, 286.2, and 288.9 eV can be observed in all three samples. The main C 1s corresponds to adventitious carbon species presenting a band located at 284.4 eV [44]. The small shoulder at 286.4 and 288.9 eV could be accounted for with the C-N-C and C-(N)$_3$ groups of g-C$_3$N$_4$, respectively [26,45,46]. In addition, the regional spectrum of N 1s for 0.05Fe-CT as seen in Figure 6d could be fitted into two peaks at 399.2 and 400.4 eV, with the former ascribed to C-N=C [34] and the latter to the N-(C)$_3$ of the g-C$_3$N$_4$ [26]. No peak concerning chemical interaction between Ti and C (Ti-C) or N (Ti-N) is seen for the 0.05Fe-CT sample in the XPS (Figure 6c,d) spectra. Taking into account of the results of FTIR, SEM, and DRS studies together, the deposition of g-C$_3$N$_4$ could only be on the surface of the Fe-TiO$_2$, and there was a chemical reaction between them as no apparent characteristic Ti-C(N) coordination peaks were seen.

2.2. Photocatalytic Activity Test

The photocatalytic activity of the prepared samples was evaluated in terms of photodegradation of phenol with a concentration of 10 mg dm^{-3} under visible light irradiation. Phenol was chosen as a model pollutant because it is the basic molecule of phenolic compounds, which are known to be highly toxic, persistent, and biorecalcitrant, widely used in preservative, herbicide, and pesticide products, and being considered to be a grave threat to the health of humankind [47]. The study was carried out under similar experimental conditions using the respective photocatalytic materials.

The change of concentration of phenol during the photodegradation process under visible light irradiation is shown in Figure 7a. The decomposition of phenol was achieved to an extent of 38.6% and 72% at 80 min for pure g-C$_3$N$_4$ and TiO$_2$, respectively. The effective photodegradation of phenol observed (79.2%) in the case of g-C$_3$N$_4$ (CT) could account for the impact of the hybrid structure on enhancing the photocatalytic activity of the material. Upon the incorporation of Fe^{3+} into CT, i.e., xFe-CT, the observed photocatalytic activity was higher compared to that of TiO$_2$ and CT, separately. Furthermore, the photocatalytic activity of the xFe-CT increased as the content of Fe^{3+} increased initially from 0.01% to 0.05%, and thereupon there was a decline trend up to 0.06%. Additionally, the 0.05Fe-CT sample showed a highest photocatalytic activity of complete degradation of phenol at 80 min under visible light irradiation. These findings confirmed that the enhanced photocatalytic activity of the composite materials cause a synergetic effect of both g-C_3N_4 and Fe^{3+}.

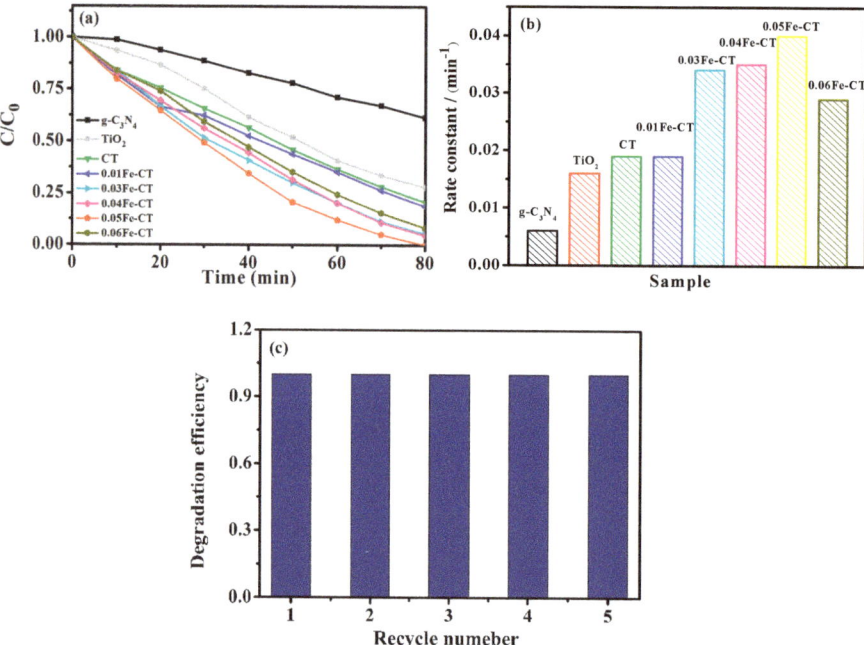

Figure 7. (a) Change of concentration of phenol during the photodegradation process; (b) the kinetic constants for all samples; (c) the cyclic test of 0.05Fe-CT for its stability.

The degradation of phenol follows a pseudo-first-order reaction [28,48] of $ln(\frac{c_t}{c_0}) = -kt$, where C_0 is the initial concentration of phenol, C_t is the concentration of phenol at time t, and k is kinetic constant. The kinetic constants (k) of all samples were calculated and are given in Figure 7b. It is 0.019 min^{-1} for CT composites higher than that of either pure g-C_3N_4 (the k value was 0.0044 min^{-1}) or TiO_2 (the k value was 0.016 min^{-1}). When Fe^{3+} ions were introduced, an enhanced photocatalytic activity was observed from all of the g-C_3N_4/Fe-TiO_2 composites and the rate constants were 0.016 min^{-1}, 0.034 min^{-1}, 0.035 min^{-1}, 0.04 min^{-1}, and 0.029 min^{-1} for the samples of 0.01Fe-CT–0.06Fe-CT, respectively. It was found that the incorporation of Fe^{3+} enhanced the photocatalytic activity of the obtained heterojunctions, and the highest performance was observed in 0.05Fe-CT.

The stability of 0.05Fe-CT was examined by catalyst recycling experiments under similar operating conditions for the photodegradation of phenol. After each cycle, the catalyst was separated by centrifugation, washed with ethanol and millipore deionized water, then dried and reused for a fresh run of photodegradation of phenol with a concentration of 10 mg dm^{-3}. As seen in Figure 7c, no obvious decline in the degradation efficiency was observed after five cycles, suggesting that the combination of g-C_3N_4 and Fe-TiO_2 has high-level photocatalytic stability for phenolic compound degradation.

2.3. Photocatalytic Mechanism

In order to elucidate the photocatalytic mechanism of Fe-CT composites, the main active species generated over g-C$_3$N$_4$, CT, 0.05Fe-CT, and TiO$_2$ were quantified by adding a suitable scavenger during the photocatalytic degradation of phenol. The reactive species corresponding to both g-C$_3$N$_4$ and Fe-TiO$_2$ were nullified as references, so that the reactive species of 0.05Fe-CT alone could be figured out. The scavengers used in this study were sodium oxalate (OA, 0.5 mmol dm^{-3}), p-benzoquinone (BQ, 0.5 mmol dm^{-3}), and isopropanol (IPA, 1 mmol dm^{-3}) against photogenerated holes (h$^+$) [49], superoxide anion radicals (•O$_2^-$) [18], and hydroxyl radicals (•OH) [49], respectively. It needs to be mentioned that the applied concentration of each scavenger did not cause any removal of phenol in the respective control experiment [23].

As shown in Figure 8, for pure g-C$_3$N$_4$, the addition of OA caused a slight decrease of the photocatalytic efficiency from 38.6% to 37.1% at 80 min, which indicates that h$^+$ was not the major reactive specie; the introduction of IPA caused a decrease to 32.5%, which indicates that •OH made a considerable contribution towards the photocatalytic degradation. When BQ was added into the reaction solution, the degradation efficiencies of phenol showed a significant fall to 10.1%. From these results, it is very clear that •O$_2^-$ is the major reactive specie in the photocatalytic reaction of pure g-C$_3$N$_4$. The order of influence was •O$_2^-$ > •OH > h$^+$.

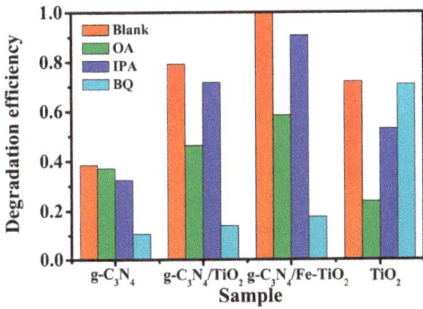

Figure 8. Trapping of reactive specie experiments for g-C$_3$N$_4$, CT, 0.05Fe-CT, and TiO$_2$.

For the TiO$_2$, the efficiency of phenol degradation was shown to be 72.1% when no scavenger was added. With the addition of OA and IPA into the reaction solution of a separate run, the photocatalytic efficiency of phenol degradation decreased to 23.8% and 52.9%, respectively. In the presence of BQ, the degradation rate of phenol was slightly decreased to 71%. Obviously, the major reactive species for pure TiO$_2$ are h$^+$ and •OH.

As seen in Figure 8, for the g-C$_3$N$_4$/TiO$_2$ photocatalytic system, the degradation efficiency of phenol was inhibited in the order BQ > OA > IPA when these three scavengers were added in the separate run, which indicates that •O$_2^-$, •OH, and h$^+$ were all of the active species generated in the g-C$_3$N$_4$/TiO$_2$ photocatalytic system.

It is clear in Figure 8 that the photocatalytic efficiency of phenol for the 0.05Fe-CT photocatalyst was 100% at 80 min without any scavengers. With the addition of scavenger IPA, BQ, and OA in the separate run, the photocatalytic efficiencies of phenol decreased to 90.1%, 58.5%, and 17.6%, respectively. The inference is that both g-C$_3$N$_4$/TiO$_2$ and Fe-CT showed an identical trend in the presence of scavengers, the major reactive species were •O$_2^-$ as well as h$^+$ in the photocatalytic reaction of g-C$_3$N$_4$/Fe-TiO$_2$, and the order of influence was •O$_2^-$ > h$^+$ > •OH.

To further determine the photocatalytic mechanism of Fe-CT composites, a quantitative estimation of •OH was carried out by the photoluminescence (PL) method using terephthalic acid (TA) as a probe molecule during the photocatalysis process. The PL signals of g-C$_3$N$_4$, TiO$_2$, Fe-TiO$_2$, CT, and 0.05Fe-CT samples recorded at 80 min of the photocatalysis process are shown in Figure 9a. It could be easily understood that no •OH was generated during the photocatalysis process using g-C$_3$N$_4$ as there was no corresponding PL signal. The absence of •OH radicals in the photocatalysis process of g-C$_3$N$_4$ could be well-explained by taking into account the position of the VB edges of g-C$_3$N$_4$ and the actual potential of the OH$^-$/•OH couple (+1.83 V/+2.7 V) (versus NHE) formation. Thus, the photogenerated holes on the surface of g-C$_3$N$_4$ were not strong enough to oxidize the OH$^-$ or H$_2$O into •OH [26,50]. However, the formation of •OH was observed in the TiO$_2$, Fe-TiO$_2$, CT, and 0.05Fe-CT samples, among which 0.05Fe-CT showed the greatest quantity of •OH generation, which confirms the Z-scheme of transferring photoexcited charge carriers between g-C$_3$N$_4$ and Fe-TiO$_2$. Otherwise, if 0.05Fe-CT worked only under the general heterojunction system, the oxidation ability of the composite would have been the same as that of g-C$_3$N$_4$, wherein the production of •OH is not possible.

As seen in Figure 9b, the linear potential part of the Mott–Schottky plot based on impedance measurements was used to determine the flat-band positions of the samples [51]. The positive slope of the straight lines indicates that both TiO$_2$ and Fe-TiO$_2$ are n-type semiconductors, i.e., the flat-band potential [52] of the samples approximately equates to the lowest potential of the CB. Thus, the CB level of TiO$_2$ and Fe-TiO$_2$ are measured to be ca +0.05 V (versus NHE) and −0.01 V, respectively. The negative shift of flat-band potentials (E_{fb}) after Fe doping suggests a similar shift of the Fermi level, which facilitates the charge separation at the semiconductor/electrolyte interface [53].

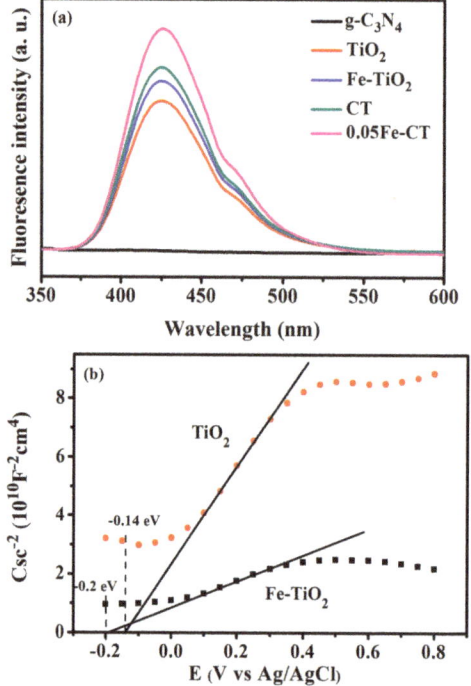

Figure 9. (a) Photoluminescence (PL) spectra of g-C$_3$N$_4$, CT, Fe-TiO$_2$, and 0.05Fe-CT in a 1 × 10^{-3} mol dm^{-3} basic solution of terephthalic acid under visible light irradiation after 80 min; (b) The Mott–Schottky plots of TiO$_2$ and Fe-TiO$_2$ for determining the flat-band potentials of samples.

These accumulated electrons in the CB of TiO$_2$ (Figure 10a) could not effectively reduce the O$_2$ to yield •O$_2^-$ due to its CB being less negative than that of •O$_2^-$/O$_2$ potential (−0.28 V versus NHE) [54]. The VB of TiO$_2$ is more positive than that of H$_2$O/•OH potential (+2.7 V versus NHE) and capable enough to oxidize H$_2$O to form •OH [34]. Our experiment mentioned above has also confirmed that h$^+$ but not the •O$_2^-$ generated was the major reactive specie in the photocatalytic degradation of the phenol molecule for pure TiO$_2$; on the other hand, the VB levels (ca. +1.58 V) of g-C$_3$N$_4$ are not positive enough to drive the oxidation of H$_2$O to form •OH, but its CB level (ca. −1.12 V) is negative enough to reduce O$_2$ to produce •O$_2^-$ [18,49]. Also, it has been observed that the •O$_2^-$ was a major reactive specie in the photocatalytic reaction for pure g-C$_3$N$_4$. For the composite of CT, the formation of •OH and the significant contribution of both h$^+$ and •O$_2^-$ in the photocatalytic reaction showed that the composite followed the Z-scheme. If the charge carriers of the CT were transferred as per the so-called usual model, the electrons in the CB of g-C$_3$N$_4$ would have migrated to the CB of TiO$_2$ and accumulated over there, which could possibly not reduce the O$_2$ to yield •O$_2^-$; holes in the VB of TiO$_2$ would migrate to the VB of g-C$_3$N$_4$, which could not oxidize $^-$OH/H$_2$O to give •OH.

For the Fe-doped TiO$_2$ (Fe-TiO$_2$), a prominent decrease in the band gap and a red shift of the threshold absorption were observed in UV-vis DRS analysis. In addition, the extent of doping of Fe^{3+}, which actually existed in the form of O$^{••}$$_v$ in the band gap of TiO$_2$, could enhance the photocatalytic activity of the material in the visible region [55,56]. As a result, the Fe-TiO$_2$ showed higher photocatalytic activity and a greater quantity of •OH generation than those of TiO$_2$.

Based on the above results, the Z-scheme mechanism of the g-C$_3$N$_4$/Fe-TiO$_2$ composites is illustrated in Figure 10b in detail. Due to their narrow band gaps, both g-C$_3$N$_4$ and Fe-TiO$_2$ can be easily excited to yield photogenerated electron–hole pairs under visible-light irradiation. Since both the CB and VB positions of Fe-TiO$_2$ are lower than those of g-C$_3$N$_4$, the photogenerated electrons (e$^-$) in the CB of Fe-TiO$_2$ tend to transfer and recombine with the photogenerated holes (h$^+$) in the VB of g-C$_3$N$_4$. The photogenerated holes left behind in the VB of Fe-TiO$_2$ can directly oxidize phenol into harmless metabolite products. Simultaneously, the remaining photogenerated electrons in the CB of g-C$_3$N$_4$ can reduce the adsorbed O$_2$ to yield •O$_2^-$, which is again a powerful oxidative species for phenol degradation. The g-C$_3$N$_4$/Fe-TiO$_2$ composites following a Z-scheme mechanism enable a fast separation and transfer of the photogenerated electron−hole pairs and in turn show strong oxidation and reduction abilities for the efficient photocatalytic degradation of organic pollutants.

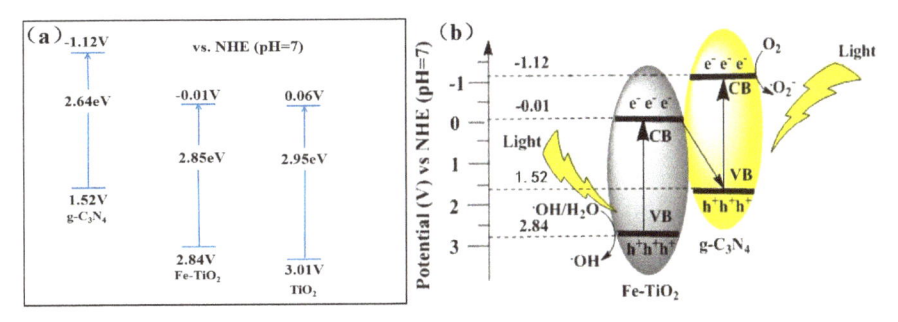

Figure 10. (a) Electronic band structure of the respective catalysts; (b) Z-scheme photocatalytic mechanism for g-C$_3$N$_4$/Fe-TiO$_2$ composites.

3. Materials and Methods

3.1. Chemical and Material

Analytical grade (AR) chemicals viz. Ferric nitrate (Fe(NO$_3$)$_3$·9H$_2$O), tetra-butyl titanate (TBOT), absolute ethyl alcohol (C$_2$H$_6$O), nitric acid (HNO$_3$), melamine, isopropanol, 5% Nafion, phenol sodium oxalate (OA), p-benzoquinone (BQ), and isopropanol were purchased from Sinopharm Chemical Reagent CO. Ltd., Shanghai, China and used as received. Millipore deionized water was used for preparing the stock solutions and the entire experimental part.

3.2. Catalyst Preparation

The g-C_3N_4 was synthesized from melamine by a direct heating step. Five grams (5 g) of melamine powder, taken in an alumina crucible, was placed in a muffle furnace and heated at 500 °C for 2 h. After cooling down to room temperature, the yellowish product was ground into powder form and again heated in a muffle furnace at 500 °C for another 2 h.

The composite particles were synthesized through a one-step hydrothermal process. A 20 mL volume of TBOT was gradually dropped into a mixture containing 167.5 mL of C_2H_6O, 5.0 mL specific concentration of $Fe(NO_3)_3$, 1.25 mL of HNO_3, and 0.047 g of g-C_3N_4 under vigorous stirring, and the stirring process continued for another 1 h. Then, the mixture was transferred into a 500 mL teflon-lined stainless steel autoclave vessel and it was kept at 200 °C for 6 h. After the hydrothermal process, the precipitate was centrifuged, washed several times with ethanol and water, dried at 80 °C overnight, and ground well. The as-prepared samples were denoted as xFe-CT, where x stands for the weight percentage of Fe (x = 0.01, 0.03, 0.04, 0.05, 0.06) with respect to TiO_2 content, and CT denotes the g-C_3N_4/TiO_2.

3.3. Characterization

The phase purity and crystal structure of the as-obtained samples were examined by the XRD technique using Rigaku Ultima IV X-ray diffraction (Rigaku Corporation, Tokyo, Japan) equipped with Cu $K\alpha$ radiation (40 kV, λ = 1.5406 Å). The 2θ scanning angle range was 20–80° with a step of 0.05 s^{-1}. The morphology was examined using a field emission scanning electron microscope (FE-SEM, NANOSEM 450, FEI Corporation, Eindhoven, the Netherlands) operating at an accelerating voltage of 30 kV. TEM characterizations were done using an H-7000FA microscope (Hitachi, Tokyo, Japan) operating at the accelerating voltage of 75 kV. The UV-visible spectrum was obtained on a UV-2550 UV-visible spectrophotometer (Shimadzu Corporation, Kyoto, Japan) at room temperature and the spectrum range analyzed was 200–800 nm. The infrared absorption spectra were measured in a frequency range from 500 cm^{-1} to 4000 cm^{-1} on a Bruker V-70 FTIR spectrophotometer (Bruker, Karlsruhe, Germany). The X-band electron paramagnetic resonance (EPR) spectra were recorded at room temperature using a Bruker A300-10/12 EPR spectrophotometer (Bruker Corporation, Karlsruhe, Germany). The microwave frequency was fixed at 100 KHz, the power was 10 mW, and the field modulation ranged between 1.3–1.9 G and 3.2–3.7 G. The X-ray photoelectron spectroscopy (XPS) data were collected by an Axis Ultra instrument (Kratos Analytical, Manchester, UK) under an ultra-high vacuum (<10^{-8} Torr) using a monochromatic Al Ka X-ray source (hv = 1486.6 eV) operating at 150 W.

Mott–Schottky plots have been investigated on an electrochemical workstation (CS310, CorrTest, Wuhan, China) under a three-electrode configuration by employing TiO_2 or Fe-TiO_2, Ag/AgCl, and Pt mesh as the working, reference, and counter electrode, respectively. Herein, the working electrodes had undergone a two-step treatment. Initially, 5 mg of synthesized sample was mixed with 800 μL of isopropanol, followed by 200 μL of millipore deionized water, and finally 20 μL of 5% Nafion, and then the mixture underwent an ultrasonication treatment; 6 μL of the mixture was dropped onto glassy carbon electrode and dried in an air environment. The supporting electrolyte used was Na_2SO_4 with a concentration of 0.5 mol dm^{-3}. The potential scanning measurements for the electrode were performed from −0.2 V to 0.8 V in dark conditions, and the impedance-potential characteristics of the electrode were recorded at a frequency of 10 Hz.

3.4. Photocatalytic Activity Measurement

The photocatalytic experiment on phenol degradation under visible light was carried out in a glass container having a volume capacity of 200 mL to evaluate the activity of the g-C_3N_4/Fe-TiO_2 composites. The light source was a 300 W PLS-SXE 300 xenon lamp (Perfect light, Wuhan, China) with a 400 nm cut filter to remove the UV irradiation that was suspended over a height of 10 cm from the reaction solution surface. Typically, 5 mg of as-prepared photocatalyst was added into 50 mL of phenol-contaminated (10 mg dm^{-3}) working solution. The glass container was placed in an ice-water bath, and the entire setup was placed on a magnetic stirrer operated at a constant stirring rate of 380 rpm. Prior to light irradiation, the suspension was stirred for 1 h to establish an adsorption/desorption equilibrium between phenol and photocatalyst under dark conditions. After visible light irradiation for a defined period of time (every 10 min), the reaction solution (1 mL) for analysis was siphoned out, and then the suspensions were removed by centrifugation and the clear supernatant solution was used for analysis. The concentration of phenol was measured by high performance liquid chromatography (HPLC) (Shimadzu, Kyoto, Japan) equipped with a UV detector and a C_{18} reverse-phase column (4.6 mm i.d. ×150 mm, Agilent, CA, USA). The mobile phase used in the HPLC was water and methanol (volumetric ratio of 50:50), and the injection volume of the sample was 20 μL with a flow rate of 0.5 mL min^{-1}. The wavelength of the UV absorbance detector was fixed at 270 nm.

The quantity of •OH in the photocatalytic process was determined by the photoluminescence (PL) technique using terephthalic acid as a probe molecule. Terephthalic acid reacts with •OH to produce a highly fluorescent product 2-hydroxy terephthalic rapidly and specifically, which reflects as the PL signal at 425 nm excited by 315 nm of light. Detailed experimental information is given in our previous work [57].

4. Conclusions

The photocatalytic g-C_3N_4/Fe-TiO_2 composite was successfully synthesized by a one-step hydrothermal process and found to exhibit excellent photocatalytic activity and stability for the photocatalytic degradation of organic pollutants. The composite with an optimum content of Fe^{3+} of 0.05 wt % exhibits the highest photocatalytic activity towards phenol degradation owing to a stronger spectral absorption of visible light wavelengths, an enhancement in the carrier density, and a decrease in the charge transfer resistance between the interface of solid and electrolyte. The formation of Z-scheme g-C_3N_4/Fe-TiO_2 heterojunctions possesses a higher efficiency of charge separation and transfer as well as stronger oxidation and reduction abilities. This work may give new insight into the development of Z-scheme composite photocatalysts, which is of a great interest to the scientific community for photocatalysis.

Acknowledgments: This work was supported by the International Science & Technology Cooperation Program of China (Nos. 2013DFG50150 and 2016YFE0126300) and the Innovative and Interdisciplinary Team at HUST (2015ZDTD027). The authors thank the Analytical and Testing Center of HUST for the use of SEM, XRD, TEM, FTIR, and DRS equipment.

Author Contributions: Yanrong Zhang and Muthu Murugananthan conceived and designed the experiments; Jie Gu performed the experiments; and Zedong Zhu contributed analysis tools and wrote the paper.

Conflicts of Interest: The authors declare no conflict of interest.

References

1. Asahi, R.; Morikawa, T.; Ohwaki, T.; Aoki, K.; Taga, Y. Visible-Light Photocatalysis in Nitrogen-Doped Titanium Oxides. *Science* **2001**, *293*, 269–271. [CrossRef] [PubMed]
2. Tong, H.; Ouyang, S.; Bi, Y.; Umezawa, N.; Oshikiri, M.; Ye, J. Nano-photocatalytic materials: Possibilities and challenges. *Adv. Mater.* **2012**, *24*, 229–251. [CrossRef] [PubMed]
3. Wang, G.; Wang, H.; Ling, Y.; Tang, Y.; Yang, X.; Fitzmorris, R.C.; Wang, C.; Zhang, J.Z.; Li, Y. Hydrogen-treated TiO_2 nanowire arrays for photoelectrochemical water splitting. *Nano Lett.* **2011**, *11*, 3026–3033. [CrossRef] [PubMed]
4. Zhou, W.; Liu, Q.; Zhu, Z.; Zhang, J. Preparation and properties of vanadium-doped TiO_2 photocatalysts. *J Phys. D Appl. Phys.* **2010**, *43*, 035301. [CrossRef]

5. George, S.; Pokhrel, S.; Ji, Z.; Henderson, B.L.; Xia, T.; Li, L.; Zink, J.I.; Nel, A.E.; Madler, L. Role of Fe doping in tuning the band gap of TiO_2 for the photo-oxidation-induced cytotoxicity paradigm. *J. Am. Chem. Soc.* **2011**, *133*, 11270–11278. [CrossRef] [PubMed]
6. Colón, G.; Maicu, M.; Hidalgo, M.C.; Navío, J.A. Cu-doped TiO_2 systems with improved photocatalytic activity. *Appl. Catal. B Environ.* **2006**, *67*, 41–51. [CrossRef]
7. Ohno, T.; Murakami, N.; Tsubota, T.; Nishimura, H. Development of metal cation compound-loaded S-doped TiO_2 photocatalysts having a rutile phase under visible light. *Appl. Catal. A Gen.* **2008**, *349*, 70–75. [CrossRef]
8. Samiolo, L.; Valigi, M.; Gazzoli, D.; Amadelli, R. Photo-electro catalytic oxidation of aromatic alcohols on visible light-absorbing nitrogen-doped TiO_2. *Electrochim. Acta* **2010**, *55*, 7788–7795. [CrossRef]
9. Yu, J.; Qi, L.; Jaroniec, M. Hydrogen Production by Photocatalytic Water Splitting over Pt/TiO_2 Nanosheets with Exposed (001) Facets. *J. Phys. Chem. C* **2010**, *114*, 13118–13125. [CrossRef]
10. Seh, Z.W.; Liu, S.; Low, M.; Zhang, S.Y.; Liu, Z.; Mlayah, A.; Han, M.Y. Janus $Au-TiO_2$ photocatalysts with strong localization of plasmonic near-fields for efficient visible-light hydrogen generation. *Adv. Mater.* **2012**, *24*, 2310–2314. [CrossRef] [PubMed]
11. Choi, W.; Termin, A.; Hoffmann, M.R. The Role of Metal Ion Dopants in Quantum-Sized TiO_2: Correlation between Photoreactivity and Charge Carrier Recombination Dynamics. *J. Phys. Chem.* **1994**, *98*, 13669–13679. [CrossRef]
12. Dai, K.; Lu, L.; Liang, C.; Liu, Q.; Zhu, G. Heterojunction of facet coupled $g-C_3N_4$/surface-fluorinated TiO_2 nanosheets for organic pollutants degradation under visible LED light irradiation. *Appl. Catal. B Environ.* **2014**, *156–157*, 331–340. [CrossRef]
13. Takahashi, Y.; Tatsuma, T. Oxidative Energy Storage Ability of a $TiO_2-Ni(OH)_2$ Bilayer Photocatalyst. *Langmuir* **2005**, *21*, 12357–12361. [CrossRef] [PubMed]
14. Su, J.; Guo, L.; Bao, N.; Grimes, C.A. Nanostructured $WO_3/BiVO_4$ heterojunction films for efficient photoelectrochemical water splitting. *Nano Lett.* **2011**, *11*, 1928–1933. [CrossRef] [PubMed]
15. Zhang, Y.C.; Yao, L.; Zhang, G.; Dionysiou, D.D.; Li, J.; Du, X. One-step hydrothermal synthesis of high-performance visible-light-driven SnS_2/SnO_2 nanoheterojunction photocatalyst for the reduction of aqueous Cr(VI). *Appl. Catal. B Environ.* **2014**, *144*, 730–738. [CrossRef]
16. Dai, X.; Xie, M.; Meng, S.; Fu, X.; Chen, S. Coupled systems for selective oxidation of aromatic alcohols to aldehydes and reduction of nitrobenzene into aniline using $CdS/g-C_3N_4$ photocatalyst under visible light irradiation. *Appl. Catal. B Environ.* **2014**, *158–159*, 382–390. [CrossRef]
17. Arai, T.; Yanagida, M.; Konishi, Y.; Iwasaki, Y.; Sugihara, H.; Sayama, K. Efficient Complete Oxidation of Acetaldehyde into CO_2 over $CuBi_2O_4/WO_3$ CompositePhotocatalyst under Visible and UV Light Irradiation. *J. Phys. Chem. C* **2007**, *111*, 7574–7577. [CrossRef]
18. Yang, Y.; Guo, W.; Guo, Y.; Zhao, Y.; Yuan, X.; Guo, Y. Fabrication of Z-scheme plasmonic photocatalyst $Ag@AgBr/g-C_3N_4$ with enhanced visible-light photocatalytic activity. *J. Hazard. Mater.* **2014**, *271*, 150–159. [CrossRef] [PubMed]
19. Wang, X.; Maeda, K.; Thomas, A.; Takanabe, K.; Xin, G.; Carlsson, J.M.; Domen, K.; Antonietti, M. A metal-free polymeric photocatalyst for hydrogen production from water under visible light. *Nat. Mater.* **2009**, *8*, 76–80. [CrossRef] [PubMed]
20. Dong, G.; Zhang, Y.; Pan, Q.; Qiu, J. A fantastic graphitic carbon nitride ($g-C_3N_4$) material: Electronic structure, photocatalytic and photoelectronic properties. *J. Photochem. Photobiol. C Photochem. Rev.* **2014**, *20*, 33–50. [CrossRef]
21. Li, K.; Gao, S.; Wang, Q.; Xu, H.; Wang, Z.; Huang, B.; Dai, Y.; Lu, J. In-Situ-Reduced Synthesis of Ti^{3+} Self-Doped $TiO_2/g-C_3N_4$ Heterojunctions with High Photocatalytic Performance under LED Light Irradiation. *ACS Appl. Mater. Interfaces* **2015**, *7*, 9023–9030. [CrossRef] [PubMed]
22. Yang, N.; Li, G.; Wang, W.; Yang, X.; Zhang, W.F. Photophysical and enhanced daylight photocatalytic properties of N-doped $TiO_2/g-C_3N_4$ composites. *J. Phys. Chem. Solids* **2011**, *72*, 1319–1324. [CrossRef]
23. Pany, S.; Parida, K.M. A facile in situ approach to fabricate $N,S-TiO_2/g-C_3N_4$ nanocomposite with excellent activity for visible light induced water splitting for hydrogen evolution. *Phys. Chem. Chem. Phys.* **2015**, *17*, 8070–8077. [CrossRef] [PubMed]
24. Kondo, K.; Murakami, N.; Ye, C.; Tsubota, T.; Ohno, T. Development of highly efficient sulfur-doped TiO_2 photocatalysts hybridized with graphitic carbon nitride. *Appl. Catal. B Environ.* **2013**, *142–143*, 362–367. [CrossRef]

25. Busca, G.; Berardinelli, S.; Resini, C.; Arrighi, L. Technologies for the removal of phenol from fluid streams: A short review of recent developments. *J. Hazard. Mater.* **2013**, *160*, 265–288. [CrossRef] [PubMed]
26. Liao, W.; Muruganantham, M.; Zhang, Y. Synthesis of Z-scheme g-C_3N_4-Ti^{3+}/TiO_2 material: An efficient visible light photoelectrocatalyst for degradation of phenol. *Phys. Chem. Chem. Phys.* **2015**, *17*, 8877–8884. [CrossRef] [PubMed]
27. Yu, J.; Dai, G.; Huang, B. Fabrication and Characterization of Visible-Light-Driven Plasmonic Photocatalyst Ag/AgCl/TiO_2 Nanotube Arrays. *J. Phys. Chem. C* **2009**, *113*, 16394–16401. [CrossRef]
28. Thomas, A.; Fischer, A.; Goettmann, F.; Antonietti, M.; Müller, J.-O.; Schlögl, R.; Carlsson, J.M. Graphitic carbon nitride materials: Variation of structure and morphology and their use as metal-free catalysts. *J. Mater. Chem.* **2008**, *18*, 4893–4908. [CrossRef]
29. Zhang, Y.; Gu, J.; Muruganantham, M.; Zhang, Y. Development of novel α-Fe_2O_3/$NiTiO_3$ heterojunction nanofibers material with enhanced visible-light photocatalytic performance. *J. Alloy Comp.* **2015**, *630*, 110–116. [CrossRef]
30. Yan, S.C.; Li, Z.S.; Zou, Z.G. Photodegradation performance of g-C_3N_4 fabricated by directly heating melamine. *Langmuir* **2009**, *25*, 10397–10401. [CrossRef] [PubMed]
31. Kumar, S.; Surendar, T.; Baruah, A.; Shanker, V. Synthesis of a novel and stable g-C_3N_4–Ag_3PO_4 hybrid nanocomposite photocatalyst and study of the photocatalytic activity under visible light irradiation. *J. Mater. Chem. A* **2013**, *1*, 5333–5340. [CrossRef]
32. Zhao, Y.; Yu, D.; Zhou, H.; Tian, Y.; Yanagisawa, O. Turbostratic carbon nitride prepared by pyrolysis of melamine. *J. Mater. Sci.* **2005**, *40*, 2645–2647. [CrossRef]
33. Islam, M.J.; Reddy, D.A.; Choi, J.; Kim, T.K. Surface oxygen vacancy assisted electron transfer and shuttling for enhanced photocatalytic activity of a Z-scheme CeO_2–AgI nanocomposite. *RSC Adv.* **2016**, *6*, 19341–19350. [CrossRef]
34. Yu, J.; Wang, S.; Low, J.; Xiao, W. Enhanced photocatalytic performance of direct Z-scheme g-C_3N_4-TiO_2 photocatalysts for the decomposition of formaldehyde in air. *Phys. Chem. Chem. Phys.* **2013**, *15*, 16883–16890. [CrossRef] [PubMed]
35. Bajnóczi, É.G.; Balázs, N.; Mogyorósi, K.; Srankó, D.F.; Pap, Z.; Ambrus, Z.; Canton, S.E.; Norén, K.; Kuzmann, E.; Vértes, A.; et al. The influence of the local structure of Fe(III) on the photocatalytic activity of doped TiO_2 photocatalysts—An EXAFS, XPS and Mössbauer spectroscopic study. *Appl. Catal. B Environ.* **2011**, *103*, 232–239. [CrossRef]
36. Egerton, T.A.; Harris, E.; John Lawson, E.; Mile, B.; Rowlands, C.C. An EPR study of diffusion of iron into rutile. *Phys. Chem. Chem. Phys.* **2001**, *3*, 497–504. [CrossRef]
37. Nagaveni, K.; Hegde, M.; Madras, G. Structure and Photocatalytic Activity of $Ti_{1-x}M_xO_{2\pm\delta}$ (M = W, V, Ce, Zr, Fe, and Cu) Synthesized by Solution Combustion Method. *J. Phys. Chem. B* **2004**, *108*, 20204–20212. [CrossRef]
38. Tong, T.; Zhang, J.; Tian, B.; Chen, F.; He, D. Preparation of Fe^{3+}-doped TiO_2 catalysts by controlled hydrolysis of titanium alkoxide and study on their photocatalytic activity for methyl orange degradation. *J. Hazard. Mater.* **2008**, *155*, 572–579. [CrossRef] [PubMed]
39. Nishikawa, M.; Mitani, Y.; Nosaka, Y. Photocatalytic Reaction Mechanism of Fe(III)-Grafted TiO_2 Studied by Means of ESR Spectroscopy and Chemiluminescence Photometry. *J. Phys. Chem. C* **2012**, *116*, 14900–14907. [CrossRef]
40. Pecchi, G.; Reyes, P.; Lopez, T.; Gómez, R.; Moreno, A.; Fierro, J.; Martínez-Arias, A. Catalytic Combustion of Methane on Fe-TiO_2 Catalysts Prepared by Sol-Gel Method. *J. Sol-Gel Sci. Technol.* **2003**, *27*, 205–214. [CrossRef]
41. Ganesh, I.; Kumar, P.P.; Gupta, A.K.; Sekhar, P.S.; Radha, K.; Padmanabham, G.; Sundararajan, G. Preparation and characterization of Fe-doped TiO_2 powders for solar light response and photocatalytic applications. *Process. Appl. Ceram.* **2012**, *6*, 21–36. [CrossRef]
42. Yalçın, Y.; Kılıç, M.; Çınar, Z. Fe^{3+}-doped TiO_2: A combined experimental and computational approach to the evaluation of visible light activity. *Appl. Catal. B Environ.* **2010**, *99*, 469–477. [CrossRef]
43. Santara, B.; Giri, P.K.; Dhara, S.; Imakita, K.; Fujii, M. Oxygen vacancy-mediated enhanced ferromagnetism in undoped and Fe-doped TiO_2 nanoribbons. *J. Phys. D Appl. Phys.* **2014**, *47*, 235304. [CrossRef]
44. Miranda, C.; Mansilla, H.; Yáñez, J.; Obregón, S.; Colón, G. Improved photocatalytic activity of g-C_3N_4/TiO_2 composites prepared by a simple impregnation method. *J Photoch. Photobio. A Chem.* **2013**, *253*, 16–21. [CrossRef]

45. Zhou, S.; Liu, Y.; Li, J.; Wang, Y.; Jiang, G.; Zhao, Z.; Wang, D.; Duan, A.; Liu, J.; Wei, Y. Facile in situ synthesis of graphitic carbon nitride (g-C_3N_4)-N-TiO_2 heterojunction as an efficient photocatalyst for the selective photoreduction of CO_2 to CO. *Appl. Catal. B Environ.* **2014**, *158–159*, 20–29. [CrossRef]
46. Choi, J.; Reddy, D.A.; Islam, M.J.; Seo, B.; Joo, S.H.; Kim, T.K. Green synthesis of the reduced graphene oxide–CuI quasi-shell–core nanocomposite: A highly efficient and stable solar-light-induced catalyst for organic dye degradation in water. *Appl. Surf. Sci.* **2015**, *358*, 159–167. [CrossRef]
47. Chen, Q.; Li, J.; Li, X.; Huang, K.; Zhou, B.; Cai, W.; Shangguan, W. Visible-light responsive photocatalytic fuel cell based on WO_3/W photoanode and Cu_2O/Cu photocathode for simultaneous wastewater treatment and electricity generation. *Environ. Sci. Technol.* **2012**, *46*, 11451–114588. [CrossRef] [PubMed]
48. Chen, X.; Murugananthan, M.; Zhang, Y. Degradation of p-Nitrophenol by thermally activated persulfate in soil system. *Chem. Eng. J.* **2016**, *283*, 1357–1365. [CrossRef]
49. Chen, S.; Hu, Y.; Meng, S.; Fu, X. Study on the separation mechanisms of photogenerated electrons and holes for composite photocatalysts g-C_3N_4-WO_3. *Appl. Catal. B Environ.* **2014**, *150–151*, 564–573. [CrossRef]
50. Yang, J.; Liao, W.; Liu, Y.; Murugananthan, M.; Zhang, Y. Degradation of Rhodamine B using a Visible-light driven Photocatalytic Fuel Cell. *Electrochim. Acta* **2014**, *144*, 7–15. [CrossRef]
51. Gelderman, K.; Lee, L.; Donne, S. Flat-Band Potential of a Semiconductor: Using the Mott-Schottky Equation. *J. Chem. Educ.* **2007**, *84*, 685. [CrossRef]
52. Sakthivel, S.; Kisch, H. Daylight photocatalysis by carbon-modified titanium dioxide. *Chem. Int. Ed.* **2003**, *42*, 4908–4911. [CrossRef] [PubMed]
53. Kontos, A.I.; Likodimos, V.; Stergiopoulos, T.; Tsoukleris, D.S.; Falaras, P.; Rabias, I.; Papavassiliou, G.; Kim, D.; Kunze, J.; Schmuki, P. Self-Organized Anodic TiO_2 Nanotube Arrays Functionalized by Iron Oxide Nanoparticles. *Chem. Mater.* **2009**, *21*, 662–672. [CrossRef]
54. Xiang, Q.; Yu, J.; Wong, P.K. Quantitative characterization of hydroxyl radicals produced by various photocatalysts. *J. Colloid. Interface Sci.* **2011**, *357*, 163–167. [CrossRef] [PubMed]
55. Wu, Q.; Zheng, Q.; van de Krol, R. Creating Oxygen Vacancies as a Novel Strategy to Form Tetrahedrally Coordinated Ti^{4+} in Fe/TiO_2 Nanoparticles. *J. Phys. Chem. C* **2012**, *116*, 7219–7226. [CrossRef]
56. Wu, Q.; van de Krol, R. Selective photoreduction of nitric oxide to nitrogen by nanostructured TiO_2 photocatalysts: Role of oxygen vacancies and iron dopant. *J. Am. Chem. Soc.* **2012**, *134*, 9369–9375. [CrossRef] [PubMed]
57. Liao, W.; Zhang, Y.; Zhang, M.; Murugananthan, M.; Yoshihara, S. Photoelectrocatalytic degradation of microcystin-LR using Ag/AgCl/TiO_2 nanotube arrays electrode under visible light irradiation. *Chem. Eng. J.* **2013**, *231*, 455–463. [CrossRef]

© 2018 by the authors. Licensee MDPI, Basel, Switzerland. This article is an open access article distributed under the terms and conditions of the Creative Commons Attribution (CC BY) license (http://creativecommons.org/licenses/by/4.0/).

catalysts

Article

Highly Efficient and Visible Light Responsive Heterojunction Composites as Dual Photoelectrodes for Photocatalytic Fuel Cell

Honghui Pan [1], Wenjuan Liao [1], Na Sun [1], Muthu Murugananthan [2] and Yanrong Zhang [1,*]

[1] Environmental Science Research Institute, Huazhong University of Science and Technology, Wuhan 430074, China; honghui_pan@hust.edu.cn (H.P.); liaowenjuan@hust.edu.cn (W.L.); nasun.hust@gmail.com (N.S.)
[2] Department of Chemistry, PSG College of Technology, Peelamedu, Coimbatore 641004, India; muruga.chem@gmail.com
* Correspondence: yanrong_zhang@hust.edu.cn; Tel.: +86-027-877-9210-7802

Received: 11 November 2017; Accepted: 8 January 2018; Published: 18 January 2018

Abstract: In the present work, a novel photocatalytic fuel cell (PFC) system involving a dual heterojunction photoelectrodes, viz. polyaniline/TiO$_2$ nanotubes (PANI/TiO$_2$ NTs) photoanode and CuO/Co$_3$O$_4$ nanorods (CuO/Co$_3$O$_4$ NRs) photocathode, has been designed. Compared to TiO$_2$ NTs electrode of PFC, the present heterojunction design not only enhances the visible light absorption but also offers the higher efficiency in degrading Rhodamine B–a model organic pollutant. The study includes an evaluation of the dual performance of the photoelectrodes as well. Under visible-light irradiation of 3 mW cm^{-2}, the cell composed of the photoanode PANI/TiO$_2$ NTs and CuO/Co$_3$O$_4$ NRs photocathode forms an interior bias of +0.24 V within the PFC system. This interior bias facilitated the transfer of electrons from the photoanode to photocathode across the external circuit and combined with the holes generated therein along with a simultaneous power production. In this manner, the separation of electron/hole pair was achieved in the photoelectrodes by releasing the holes and electrons of PANI/TiO$_2$ NTs photoanode and CuO/Co$_3$O$_4$ NRs photocathode, respectively. Using this PFC system, the degradation of Rhodamine B in aqueous media was achieved to an extent of 68.5% within a reaction duration of a four-hour period besides a simultaneous power generation of 85 µA cm^{-2}.

Keywords: polyaniline; titanium dioxide; copper(II) oxide; cobalt oxide(II,III); photocatalytic fuel cell

---

1. Introduction

Water pollution, a serious issue of global concern, is no doubt a grave threat to human health and societal progress. The contamination of natural water systems is mainly due to lack of effective and viable techniques and excessive discharge of wastewater containing toxic organic contaminants. Developing an effective purification technique to maintain a green ecological environment and simultaneously recover the chemical energy stored abundantly in toxic organics that usually let out as wastewater has become an urgent need of the hour [1–3]. A novel device, the so-called photocatalytic fuel cell (PFC), constituted with a photoanode and a photocathode, for wastewater treatment along with simultaneous electricity generation under solar irradiation, is an emerging and attractive technique in the energy and environmental domain [4–6]. In this system, the electron/hole pairs can be generated from the photoelectrodes under light irradiation in a defined wavelength region. The electrons produced from the photoanode leave the holes and transfer through the external circuit to the photocathode, and the holes at the photoanode are released for degradation of organic compounds [7]. In addition, the PFC system, in which an n-type semiconductor generally used as photoanode with a Fermi level higher than that of the cathode, could develop interior bias which facilitates the transfer of electrons from photoanode to photocathode thereby producing a concurrent generation of electricity [8]. The existing PFC systems constituting TiO_2 and Pt as photoelectrodes [3,8] have severe limitations because TiO_2 responds most to ultraviolet (UV) region light and suffers from the high probability of electron/hole pair recombination [9]. Using Pt as photocathode is obviously not a viable approach, which would eventually restrict its application to a large-scale level [10].

To overcome the above mentioned drawbacks, the studies on either developing visible light responsive photoanodes [11–14] or replacing Pt by p-type semiconductor as the photocathode [15,16] become equally important, leading to the development of dual photoelectrodes for PFC system. However, so far, these PFCs have been identified with its shortcomings that the photoactivity and photostability of the electrodes are poor, which limits their application. For instance, the visible-light driven PFC system using CdS/TiO_2 or WO_3/W as photoanode and Cu_2O/Cu as photocathode suffers from poor stability of the CdS, WO_3 and Cu_2O due to their photocorrosion nature in aqueous media [17]. The limited usage of these electrodes could be attributed largely to the inherent drawbacks of the material that result in poor response in visible-light region, weak stability and undesirable photoactivity, which eventually limit the performance of the PFC system.

Forming a heterojunction by two different semconductors is an effective strategy to facilitate the hole/electron seperation and enhance the photocatalytic activity [18,19]. For instance, polyaniline (PANI), a conducting polymer, might be a good choice for TiO_2 sensitization [20] due to high absorption coefficients in the visible-light region, high mobility of charge carriers and good environmental stability. $PANI/TiO_2$ nanocomposite could be obtained by mixing commercial TiO_2 nanopowder with PANI by a chemical oxidative polymerization step [20,21]. On the other hand, oxides of copper and cobalt, which are well known for their p-type semiconducting behavior, could be used as photocathode [22–24] replacing the noble metals and also as photocatalysts for degradation of pollutants [25]. These materials can withstand in the multiple processing steps and have a compatibility nature with other material systems. All these notable characteristics behavior make them attractive and interesting base materials for heterostructures. Chopra et al. [25,26] recently established a fact that CuO nanowire–Co_3O_4 nanoparticle heterostructure has shown a unique photoactivity under visible-light irradiation. The p–p junctions formed by the combination of CuO and Co_3O_4 could efficiently reduce the probability of recombination of photogenerated electron/hole pairs, which in turn enhances the photocatalytic activity.

In this work, a pair of materials, viz. $PANI/TiO_2$ nanotubes (NTs) as photoanode and CuO/Co_3O_4 nanoparticles (NRs) as photocathode fabricated based on Ti substrate, was proposed as a novel PFC system, which exhibits an effective degradation behavior toward Rhodamine B and shows an efficient generation of electricity.

2. Results and Discussion

2.1. Characterization of $PANI/TiO_2$ NTs Photoanode

The microstructure and elements distribution of TiO_2 and $PANI/TiO_2$ NTs were studied by using Scanning Electron Microscope (SEM) and Energy Dispersive Spectrometer (EDX) (Figure 1), respectively. The TiO_2 nanotube arrays were covered by a layer of discretely adhered PANI. As seen in the inset (Figure 1a), the cross-sectional view of TiO_2 nanotube arrays substrate clearly displays the vertically oriented nanotubes with a length of about 900 nm and a wall thickness of 10 nm. Additionally, from the EDX analysis, the existence of elements, viz. C, N, Ti and O, was confirmed and especially the minimum quantity of PANI with respect to the content of TiO_2 substrate material was confirmed by the relatively low intensities observed against the elements C and N. Figure 2 shows the X-ray diffraction patterns (XRD) recorded for both TiO_2 and $PANI/TiO_2$ NTs materials. The peaks presented for $PANI/TiO_2$ NTs reflect characteristics of anatase TiO_2 and the predominant peak of 2θ at 25.2° indicates a fine preferential growth of the Titania nanotube (TNTs) in 101 orientation

(JCPDS no. 21-1272). The fact that no diffraction peak was observed for PANI might be due to its amorphous phase in the composite. The position and shape of the peaks observed in XRD patterns for PANI/TiO$_2$ NTs were almost identical with that of TiO$_2$, indicate that the incorporation of PANI has no influence in the lattice structure of TiO$_2$, which would be an added advantage for the hybrid photocatalytic material.

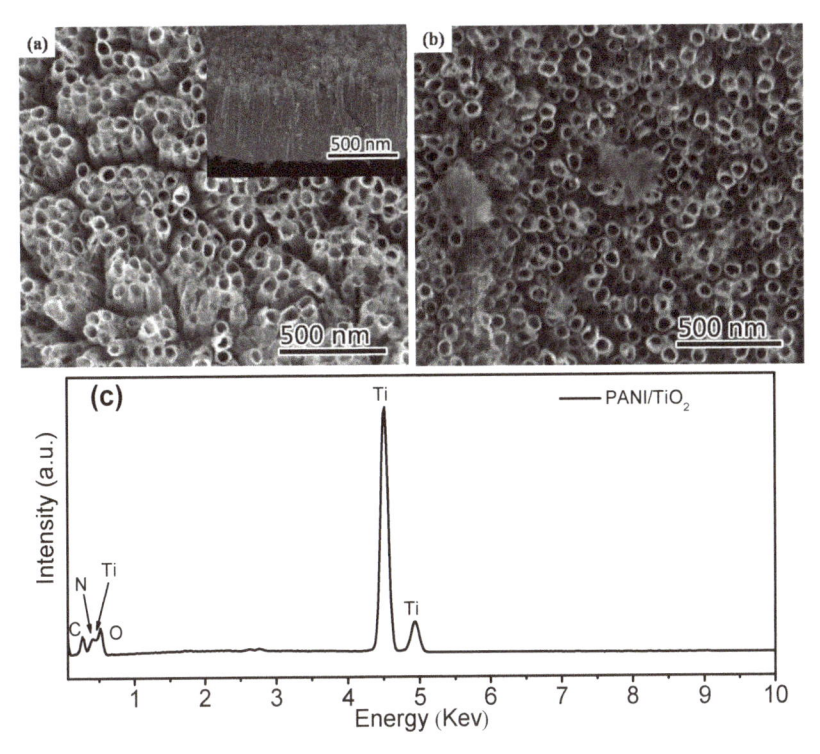

Figure 1. (a) SEM image of the TiO$_2$ (the inset shows the cross-sectional image) and (b) polyaniline (PANI)/TiO$_2$ nanotubes (NTs), (c) EDX analysis of (b).

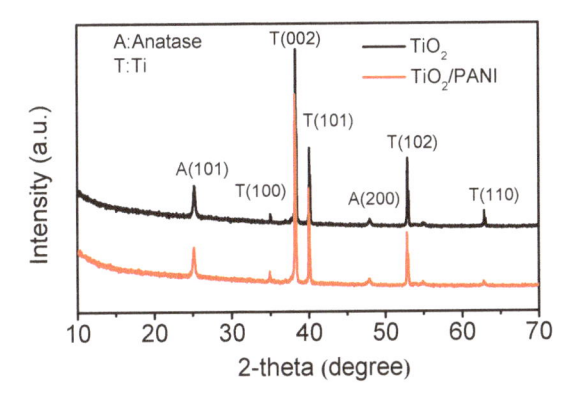

Figure 2. XRD patterns recorded for TiO$_2$, and PANI/TiO$_2$ NTs.

Figure 3 shows the Fourier Transform Infrared Spectroscopy (FTIR) spectra recorded for TiO_2 and $PANI/TiO_2$ NTs as well. The wide peak observed at 500–800 cm^{-1} for TiO_2 could be ascribed to the Ti–O bending mode of TiO_2 sample. The strong characteristic absorption bands observed for $PANI/TiO_2$ NTs, between 1200 and 1600 cm^{-1}, were 1566, 1487, 1299, 1245 and 1127 cm^{-1}. The bands at 1566 and 1487 cm^{-1} could be correlated to C–C stretching mode of quinonoid and benzenoid units, respectively. The bands at 1299 and 1245 cm^{-1} represented the C–N stretching mode of benzenoid unit while the band at 1127 cm^{-1} reflects the plane bending vibration of C=N. The bands at 796 cm^{-1} represented the C–H stretching mode of benzenoid rings [27]. Furthermore, as seen in Figure 4, the optical responses investigated by UV–vis Diffuse Reflectance Spectra (DRS) for TiO_2 and $PANI/TiO_2$ NT samples exhibit a notable absorption extension in the visible-light region at 420 nm upon the incorporation of PANI, which corresponds to a reduced bandgap absorption edge of 2.9 eV. It could be inferred from the red shift of the absorption wavelength that the $PANI/TiO_2$ NTs would be an effective visible-light driven photocatalytic material.

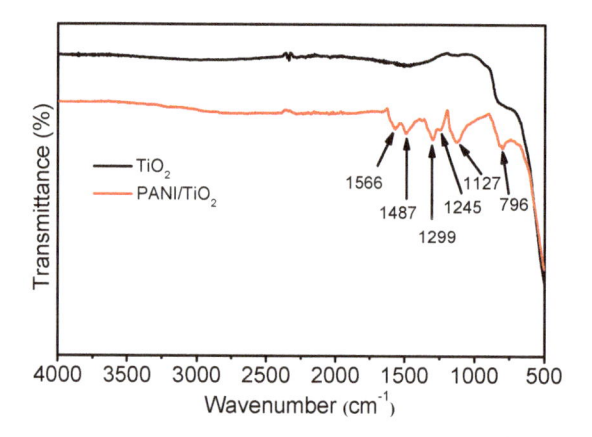

Figure 3. FTIR spectra of the TiO_2 and $PANI/TiO_2$ NTs.

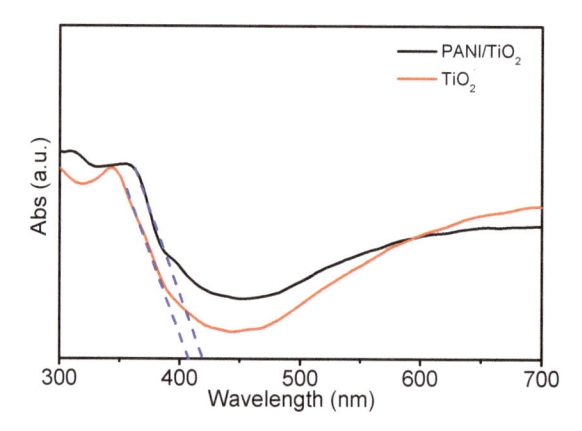

Figure 4. UV–vis diffuse reflection spectra of the TiO_2 and $PANI/TiO_2$ NTs.

In order to understand the separation and recombination of electron–hole pairs that take place in the photocatalytic materials, the photocurrent and the electrochemical impedance spectra (EIS) measurements were carried out under visible-light irradiation. The transient photocurrent responses of TiO_2 and $PANI/TiO_2$ NTs electrodes were recorded via several on–off cycles of irradiation, and the representative traces observed are shown in Figure 5a. Obviously, the intensity of photocurrent response was found to be higher for PANI/TiO$_2$ NTs (50 µA cm^{-2}) than that for TiO$_2$ NTs. The respective Nyquist plots of the TiO$_2$ and PANI/TiO$_2$ NTs photoelectrodes were shown in Figure 5b. The semicircle at high frequencies was characteristic of the charge transfer process and its diameter was equal to the charge transfer resistance. The PANI/TiO$_2$ NTs sample showed a smaller semicircle than that of TiO$_2$ sample in the Nyquist plots. This clearly confirms that the rate of electron transfer between the interface of PANI/TiO$_2$ NTs and the electrolyte was improved as a result of the deposition of PANI which causes the enhanced photoelectrochemical activity of the former compared with that of the latter.

The effects analysis of radicals was carried out in the present study to establish the PEC degradation mechanism as it is proved to be an effective approach in predicting the photodegradation reaction pathways of organic molecules that take place on the surface of the photocatalyst. The nature of interaction between the chosen scavenger and the photocatalyst makes a prominent impact on the efficiency of organic pollutant degradation. The scavengers used in this study were sodium oxalate (Na$_2$C$_2$O$_4$) of 0.5 mmol L^{-1} [28,29], isopropanol of 1 mmol L^{-1} [30], Cr(VI) of 0.05 mmol L^{-1} [28], and p-benzoquinone of 0.5 mmol L^{-1} [29,30] against h$^+$, •OH, e$^-$ and O$_2$•$^-$, respectively.

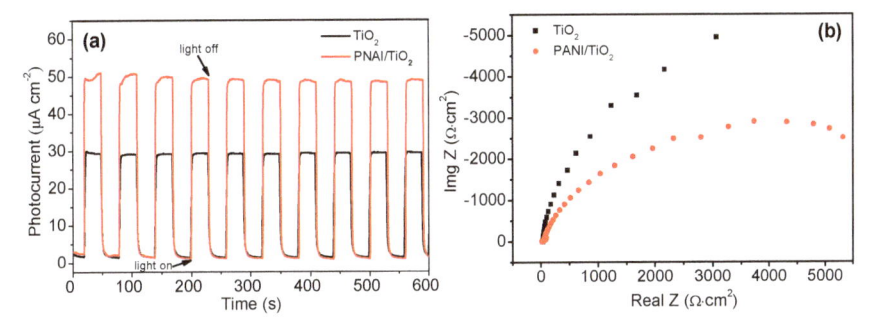

Figure 5. (a) Photocurrent responses of TiO$_2$ and PANI/TiO$_2$ NTs in 0.1 mol L^{-1} Na$_2$SO$_4$ at a bias potential of +0.6 V (vs. saturated calomel electrode (SCE)); (b) Nyquist plots of TiO$_2$ and PANI/TiO$_2$ NTs measured at open circuit potential under irradiation.

As shown in Figure 6, in the absence of a scavenger, the PEC degradation of Rhodamine B on TiO$_2$ sample at three hours was to an extent of 53%, and it was decreased to 29.4%, 35.8% and 32%, with the addition of scavengers Na$_2$C$_2$O$_4$, isopropanol and p-benzoquinone, respectively, as a separate experiment. However, in the case of Cr(VI) addition, no prominent difference was observed in the efficiency of PEC degradation (52.6%), which could be attributed to the following facts. The addition of Cr(VI) accepts photoelectron and suppresses the reduction of oxygen that results in a decreased production of O$_2$•$^-$, which in turn restrains the degradation of Rhodamine B. On the other hand, Cr(VI) inhibits the recombination of the photoinduced electron and the hole to a certain extent, which could reversely promote the efficiency of PEC degradation. Hence, the addition of Cr(VI) has no impact on the PEC degradation of Rhodamine B. It could be inferred that the major reactive species formed on pure TiO$_2$ were h$^+$, O$_2$•$^-$ and •OH.

For the PANI/TiO$_2$ NTs, under similar experimental conditions, the PEC degradation of Rhodamine B decreased from 77%, an actual efficiency obtained without scavenger, to 27.7%, 45% and 28.4% with the addition of Na$_2$C$_2$O$_4$, isopropanol and p-benzoquinone, respectively. Further, with the addition of Cr(VI) scavenger, the PEC degradation of Rhodamine B decreased to an extent of 60.4%. The results suggest that the major reactive species formed on PANI/TiO$_2$ NTs photocatalyst were e$^-$, h$^+$, •OH, and O$_2$•$^-$ with an order of influence as h$^+$ > O$_2$•$^-$ > •OH > e$^-$.

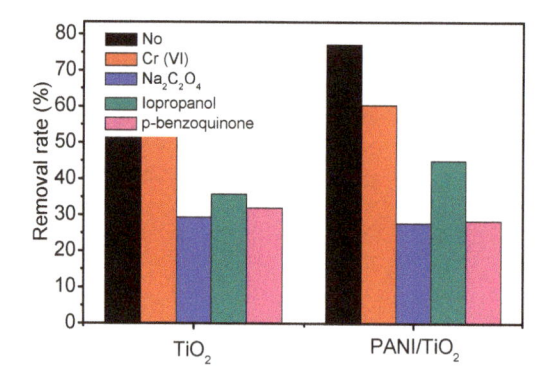

Figure 6. Effects of different scavengers on the PEC degradation of Rhodamine B (0.05 mmol L^{-1} Cr(VI): e$^-$ scavenger, 1 mmol L^{-1} isopropanol: •OH scavenger, 0.5 mmol L^{-1} p-benzoquinone: O$_2$•$^-$, 0.5 mmol L^{-1} sodium oxalate: h$^+$ scavenger).

2.2. Characterization of CuO/Co$_3$O$_4$ NRs Photocathode

The SEM images recorded for as-prepared Co$_3$O$_4$ and CuO-coated Co$_3$O$_4$ (CuO/Co$_3$O$_4$) nanorods on the Ti substrate, are shown in Figure 7a,b respectively. As seen, the diameter of the former was observed to be about 150 nm. The CuO/Co$_3$O$_4$ nanorods were fabricated by conducting 30 cycles of pulsed electrodeposition in the aqueous media containing both CuSO$_4$ and lactic acid, followed by an annealing step. As the deposition cycles increase, the CuO NPs started covering the surface of Co$_3$O$_4$ NRs (Figure 7c) gradually and upon 40 cycles, the entire surface was completely covered by CuO NPs which makes it weaken in the adsorption of incident light. The XRD pattern (Figure 8b) revealed the crystal structure and phase purity of both Co$_3$O$_4$ NRs and CuO/Co$_3$O$_4$ NR heterostructures. For Co$_3$O$_4$ NRs, all peaks in the pattern could be indexed using the Co$_3$O$_4$ anatase phase (JPCDS No: 42-1467), and the intense peak of 2θ at 19.0°, 31.2° and 36.5° could be correlated to (111), (200) and (311) plane diffractions, respectively. With the loading of CuO NPs, an additional peak of 2θ at 35.5° was observed in the (111) orientation [31]. This indicates that the deposit made on Co$_3$O$_4$ NRs was only in the form of CuO and not as Cu or Cu$_2$O. As seen in the UV–vis DRS recorded for CuO/Co$_3$O$_4$ NR sample (Figure 8b), a strong absorption was observed in the visible-light region with the band gap energy of 2.33 eV by a linear extrapolation in the absorption edge of the spectrum.

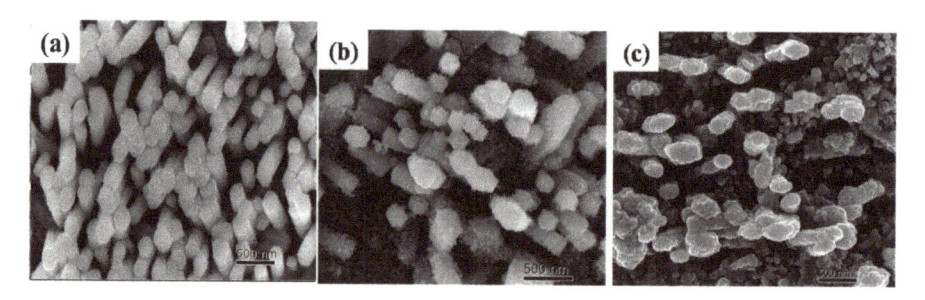

Figure 7. SEM images of the Co$_3$O$_4$ (**a**), CuO (30)/Co$_3$O$_4$ (**b**) and CuO (40)/Co$_3$O$_4$ (**c**).

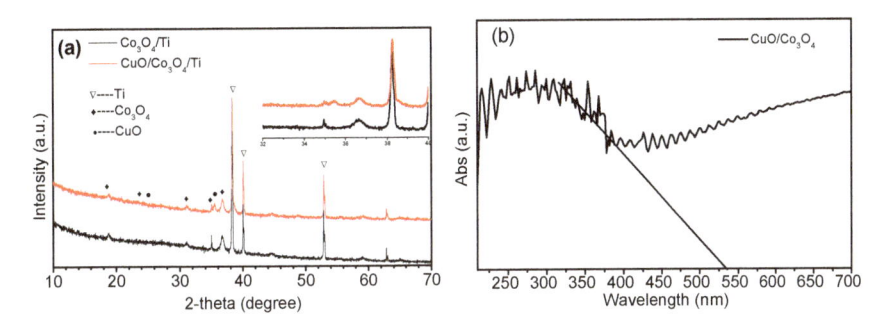

Figure 8. (a) XRD of the Co_3O_4 and CuO (30)/Co_3O_4, (b) UV–vis DRS of CuO (30)/Co_3O_4.

The influence of content of CuO NRs on the PEC performance of CuO/Co_3O_4 was studied. The NRs were fabricated by pulsed electrodeposition of different cycles, viz. 10, 20, 30 and 40. Figure 9a shows the comparative transient photocurrent response observed in applying the alternative on–off visible-light illumination cycles at −0.25 V (vs. SCE). The CuO/Co_3O_4 NRs showed an instant photoresponse under irradiation, and the photocurrent densities started increasing initially as the coating cycle increases from 10 to 30, followed by a decrease with a further increase up to 40 cycles. The maximum photocurrent density of about 170 μA cm^{-2} was observed for CuO/Co_3O_4 NRs at a coating cycle of 30. Figure 9b shows a linear sweep study for CuO/Co_3O_4 NRs processed in the potential range of −0.35 V to +0.01 V (vs. SCE) under chopped visible-light irradiation with a scan rate of 0.5 mV s^{-1}. With a cathodic potential scanning, the photocurrent was observed to be increased gradually, which is in accordance with the property of a p-type semiconductor [31]. The CuO/Co_3O_4 NRs, prepared by 30 cycles of pulsed electrodeposition was chosen as the photocathode for the PFC system of present study as it exhibits the best photoactivity.

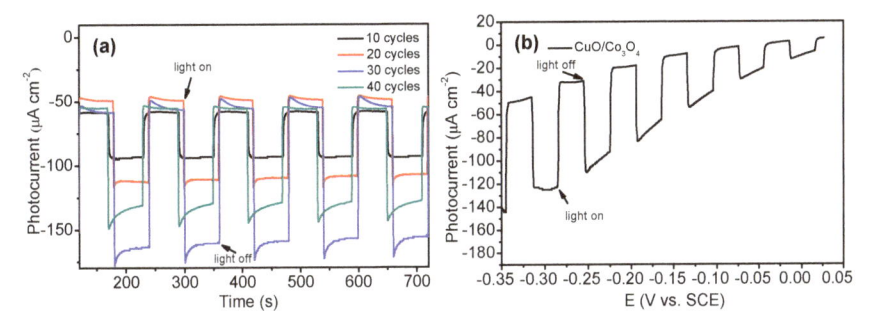

Figure 9. (a) PEC performance of the composite samples prepared at different pulse cycles at −0.25 V (vs. SCE) under visible-light irradiation in 0.1 mol L^{-1} Na_2SO_4 aqueous solution and (b) Linear sweep voltammetry (LSV) curves of CuO(30)/Co_3O_4 in 0.1 mol L^{-1} Na_2SO_4 solution in dark and under visible-light irradiation.

2.3. Characterization of PFC System and Its Performances

Figure 10 shows the Mott–Schottky (MS) plots depicted as $1/C^2$ vs. potential at 100 Hz for the respective PANI/TiO_2 NTs and CuO/Co_3O_4 NRs samples. The slopes of the linear part of the curves in the MS plot for the PANI/TiO_2 NTs were positive, which is a characteristic behavior of typical n-type semiconductor. The linear parts of the curves were x-extrapolated to zero, to obtain the V_{fb} value [32,33] of ca. −0.25 V vs. SCE for the PANI/TiO_2 NTs (Figure 10a), which represents its conduction band edge (CB). Conversely, the p-type characteristic behavior of CuO/Co_3O_4 NRs was verified by a negative slope in the MS plot, as seen in Figure 10b. The valence band (VB) edge +0.58 V vs. SCE was approximately equal to the flatband position.

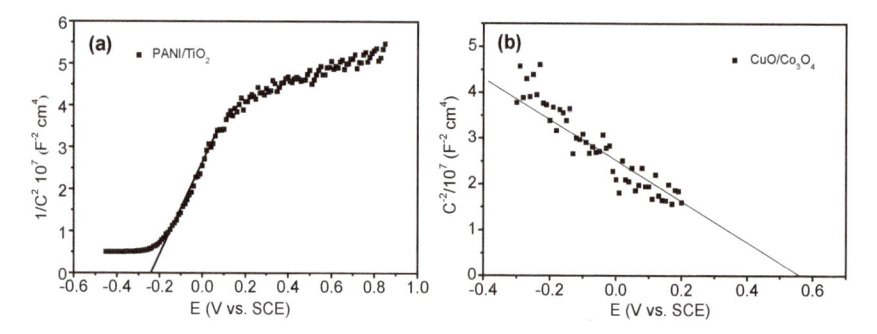

Figure 10. Mott–Schottky plots measured at a frequency of 100 Hz of (**a**) PANI/TiO$_2$ NTs, (**b**) CuO/Co$_3$O$_4$ NRs in the dark.

The energy band positions of the photoanode and photocahotde are illustrated in Figure 11a. As the Fermi level of CuO/Co$_3$O$_4$ NRs is more positive than that of PANI/TiO$_2$ NTs, an interior bias could be formed by connecting the two photoelectrodes directly, which would obviously drive the electrons generated from PANI/TiO$_2$ NTs through the external circuit and combine with the holes generated in CuO/Co$_3$O$_4$ NRs. Meanwhile, the holes and the electrons remained in the respective photoelectrode can be very well utilized for degradation of organic pollutant. It is actually the key factor that makes the PEC technique successful by combining n-type photoanode and p-type photocathode.

The open circuit potential (E_{ocp}) was established from the difference in the Fermi level of the two photoelectrodes [32,33]. To examine the photoelectric properties of the PFC, the photovoltage curves of the PFC system of PANI/TiO$_2$-CuO/Co$_3$O$_4$ was measured in the dark and under irradiation. E_{ocp} of the PANI/TiO$_2$ NTs photoanode and the CuO/Co$_3$O$_4$ NRs photocathode were measured to be −0.13 V and 0.12 V, respectively, under visible-light irradiation (3 mW cm^{-2}). It implies that the photovoltage between the photocathode and the photoanode would be +0.25 V which is consistent with the measured value (+0.24 V) of the PFC system, as shown in Figure 11b. As a result, the separation of the electron/hole pair in the photoelectrodes could be facilitated in parallel under visible-light irradiation.

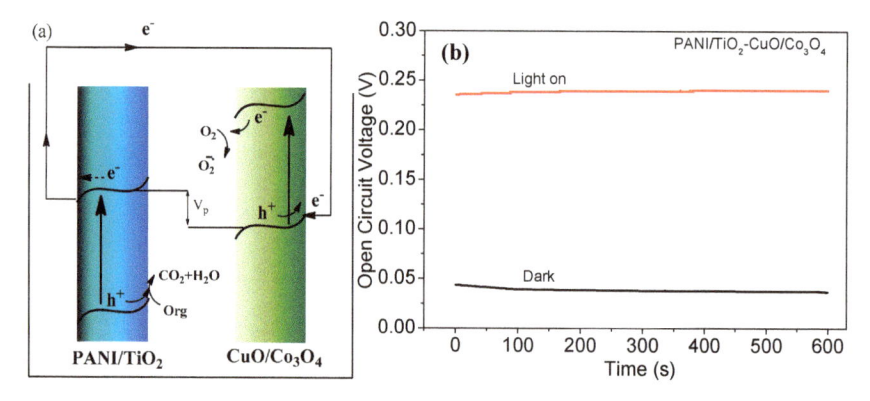

Figure 11. (**a**) Energy level diagram of the PFC cell for organic compounds degradation and electricity generation, (**b**) The open-circuit voltage of PFC cell of PANI/TiO$_2$-CuO/Co$_3$O$_4$ in dark and under visible-light irradiation.

2.4. Degradation of Rhodamine B

The performance of the PFC system was evaluated by a degradation study on Rhodamine B contaminated aqueous solution under visible-light irradiation. The degradation efficiency was monitored in terms of decolorization of Rhodamine B. The photocatalytic activity of various systems using different types of photocatalysts was compared under incandescent light irradiation as shown in Figure 12a. As seen, the photocatalytic activity of the system in which the photoelectrodes are not externally interconnected, was found to be inferior to the others and showing a decolorization of Rhodamine B of only 25.4%. For the PFC system of different photoelectrode couples TiO_2-CuO/Co_3O_4, and PANI/TiO_2-CuO/Co_3O_4 the decolorization was 51% and 68%, respectively at same reaction period. Figure 12b demonstrates that the short-circuit current density curve obtained for the present PFC system (PANI/TiO_2-CuO/Co_3O_4) during the process of Rhodamine B decolorization, was relatively steady with a current density of 85 μA cm^{-2} throughout the process. The consistent photocurrent density observed for the PFC confirmed its photostability and durability for long-time application.

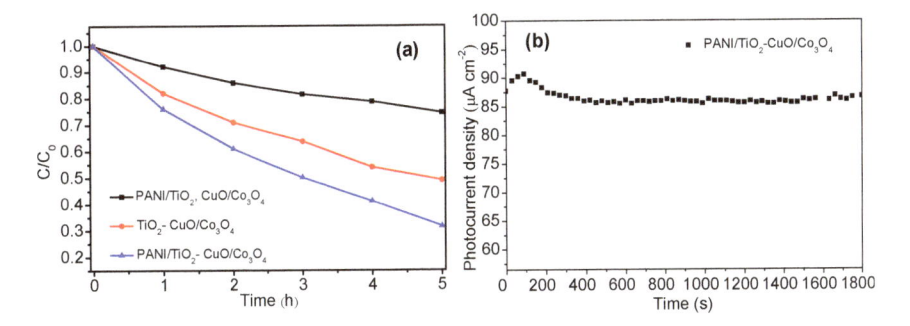

Figure 12. (a) Comparison of the degradation rates of Rhodamine B in the photocatalytic decomposition processes using unconnected PANI/TiO_2 and CuO/Co_3O_4 photoelectrodes, the PFC systems of TiO_2CuO/Co_3O_4 and PANI/TiO_2-CuO/Co_3O_4, (b) the generated electricity of the PFC.

3. Materials and Methods

3.1. Chemical and Material

Titanium foil with a thickness of 1 mm and a purity of 99.9% was purchased from Strem Chemicals (Newburyport, MA, USA). The chemicals such as ethylene glycol (EG), ammonia fluoride (NH$_4$F), sodium sulfate (Na$_2$SO$_4$), phenylamine (C$_6$H$_5$NH$_2$), cobalt nitrate hexahydrate (Co(NO$_3$)$_2$·6H$_2$O), hexamethylenetetramine (C$_6$H$_{12}$N$_4$, HMT), sodium persulfate (Na$_2$S$_2$O$_3$) and HCl were purchased from Acros Organics (Pittsburgh, PA, USA) and used as received. The aqueous solution used was prepared by using a millipore deionized (DI) water with a resistivity of 18.2 MΩ cm.

3.2. Preparation of PANI/TiO$_2$-NTs

The self-organized TiO$_2$ nanotube arrays (TiO$_2$-NTs) were fabricated on Ti foil substrate by anodization method using ethylene glycol (EG) as electrolyte media which contains 0.5 wt % NH$_4$F and 10 vol % water. The fabrication process was described in detail in our previous studies [34]. The anodization of Ti foil was performed with a two-electrode electrochemical system employing Pt mesh as cathode at a constant operating potential of 20 V for a period of 2 h. The inter electrode gap was fixed as 3 cm for every electrolysis run. In the post treatment, the anodized sample was washed with millipore deionized water, dried at 70 °C and calcined at 450 °C for 2 h.

The PANI/TiO$_2$ NTs composite was synthesized by a sequential chemical bath deposition (SCBD) method. Typically, the TiO$_2$ NTs was successively immersed into four different glass beakers for about

30 min in each beaker. The first beaker contained aqueous solution of 0.27 mol L^{-1} of phenylamine, and the third one contained an aqueous mixture of 0.23 mol L^{-1} of sodium persulfate and 0.15 mol L^{-1} of HCl, and the other two contained distilled water to rinse the samples to scavenge the excess of each precursor solution. Such an immersion treatment cycle was repeated thrice.

3.3. Preparation of CuO/Co$_3$O$_4$

The synthesis of Co$_3$O$_4$ electrode was accomplished by a simple hydrothermal process [35]. 7.2 g of Co(NO$_3$)$_2$·6H$_2$O, 0.13 g of NH$_4$F, and 0.3 g of HMT were dissolved in the order in a 50-mL acetone−deionized water (v/v = 50:50) mixture solution under continuous stirring using magnetic stirrer. Upon a formation of pink suspension, the stirring was continued for another 10 min. Then the suspension, together with a Ti film, was transformed to a teflon-lined stainless-steel autoclave vessel and kept for 24 h at 95 °C. The pink-depositions-covered-Ti film was obtained by these steps, and carefully rinsed with deionized water and dried at 70 °C, followed by a calcination process at 350 °C for 1 h in air environment. The transformation of pink depositions into black one upon calcinations confirmed the formation of Co$_3$O$_4$.

CuO was prepared by a pulsed galvanostat method under high current conditions [36]. The electrodeposition was carried out in a conventional three-electrode electrochemical workstation (CS310, CorrTest, Wuhan, China) with a conditioned electrolyte solution of 0.4 mol L^{-1} CuSO$_4$ and 3 mol L^{-1} lactic acid fixing the pH at 7 by NaOH and the temperature at 25 °C. The concentrated lactic acid acts as a complex agent for the stabilization of copper ions [37]. Upon subjecting to a negative current pulse for 0.5 s followed by a constant current density of 50 mA for 7 s, the surface of the Co$_3$O$_4$ was covered with Cu nanoparticles (NPs). The as-prepared electrode was carefully rinsed with millipore deionized water and dried at 70 °C, followed by a calcination process at 350 °C for 1 h. Then the samples were rinsed with ethanol, followed by a heat treatment at 450 °C for 1 h in air environment. In order to optimize the deposition of CuO NPs on the Co$_3$O$_4$, the samples were fabricated at different pulse cycles, viz. 10, 20, 30 and 40.

3.4. Characterization

The morphology and microstructure of the synthesized samples were characterized by field emission scanning electron microscopy (FE-SEM; NANOSEM 450, FEI, Eindhoven, The Netherlands). The phase and elemental composition of the samples were investigated using X-ray Diffraction Technique (XRD; PW3040/60 PANalytical, Almelo, The Netherlands) with Cu K α radiation (λ = 1.54056 Å). UV–visible spectrum scanning was carried out in the range of 200–800 nm using a UV-2550 model UV–visible spectrophotometer (Shimadzu Corporation, Kanagawa, Japan) at room temperature. The infrared absorption spectra were measured on a Bruker V-70 Fourier transform infrared (FTIR, Bruker, Karlsruhe, Germany) spectrophotometer in the frequency range of 400 to 4000 cm^{-1}.

The photoresponsive test was carried out for the sample (either PANI/TiO$_2$ NTs or CuO/Co$_3$O$_4$ NRs) used as working electrode in a three-electrode electrochemical work station (CS310, CorrTest, Wuhan, China), wherein saturated calomel electrode (SCE) and Pt foil was used as reference and auxiliary electrodes, respectively. The electrochemical impedance spectroscopic (EIS) studies were performed between 100 kHz and 0.01 Hz with a 5 mV rms sinusoidal modulation at the open circuit potential of the system under illumination. The linear sweep was evaluated under chopped light irradiation with a scan rate of 0.5 mV s^{-1}. Mott–Schottky plots were measured at a frequency of 100 Hz. The electrochemical studies described above were carried out in a 0.1 mol L^{-1} Na$_2$SO$_4$ aqueous solution at room temperature. The light source used was a 11 W incandescent lamp (PHILPS, Amsterdam, The Netherlands) that produced irradiation with an intensity of 3 mW cm^{-2} to the test sample which was measured by a visible-light radiometer (FZ-A, Wuhan, China).

The photoelectrochemical characteristics of the PFC were examined by connecting PANI/TiO$_2$ NTs electrode and CuO/Co$_3$O$_4$ NRs electrode directly. The short circuit current plot, and the open circuit potentials plot as well as the characteristic nature of photocurrent potentials were tested by digit precision multimeter (Tektronix DMM4050, Johnston, OH, USA) and the electrochemical workstation, respectively.

3.5. Photoelectrocatalytic Degradation of Phenol under Visible-Light Irradiation

Photoelectrocatalytic oxidation experiments were carried out in a glass container having volume capacity of 150 mL with a standard three-electrode configuration using synthesized PANI/TiO$_2$-NTs as photoanode, a Pt foil and a SCE as counter and reference electrodes, respectively. The photoelectrochemical degradation experiments were performed with a working volume of 45 mL aqueous solution containing a model contaminant Rhodamine B (1×10^{-5} mol L^{-1}) along with 0.1 mol L^{-1} Na$_2$SO$_4$ as supporting electrolyte. The glass container was placed in a water bath wherein the temperature was constantly maintained as 298 K, and the entire set-up was placed on a magnetic stirrer operated at a constant stirring rate of 650 rpm during the process. Prior to the light irradiation, the experimental solution was stirred in the dark for ca. 30 min to establish the adsorption/desorption equilibrium between the organic contaminant and the surface of the PANI/TiO$_2$-NTs under ambient air equilibrium. The degradation rate of Rhodamine B was followed by using a UV–vis spectrophotometer (UV2102 PCS, UNICO, Shanghai, China) in which the wavelength was fixed at 554 nm.

The PFC degradation of Rhodamine B (1×10^{-5} mol L^{-1}) was performed by exposing the light on both the PANI/TiO$_2$-NTs photoanode and CuO/Co$_3$O$_4$ photocathode with the illumination area of 2×2 cm^2 under similar conditions to those followed in the photoelectrocatalytic experiment. The PFC current was measured by using a digit precision multimeter.

4. Conclusions

A highly efficient and visible-light responsive photocatalytic fuel cell (PFC) system involving a dual heterojunction PANI/TiO$_2$ photoanode and CuO/Co$_3$O$_4$ photocathode was constructed. The results obtained showed that a photocurrent of 50 µA cm^{-2} was achieved using the PANI/TiO$_2$ photoanode at a bias potential of +0.6 V (vs. SCE) in 0.1 mol L^{-1} Na$_2$SO$_4$ electrolyte under visible-light irradiation of 3 mW cm^{-2}, which was 150% higher than that of TiO$_2$. Additionally, the optimized CuO/Co$_3$O$_4$ photocathode exhibited a photocurrent of 170 µA cm^{-2} at -0.25 V (vs. SCE). The PFC was constructed with the aim of providing an internal bias potential to the photoelectrocatalytic system and the performance and working mechanism of the same were systematically investigated. Under visible-light irradiation, the interior bias (+0.24 V) developed, drives the electrons of the PANI/TiO$_2$-NT's photoanode across the external circuit to combine with the holes of the CuO/Co$_3$O$_4$ photocathode, which actually leads to electron/hole pair separation at respective photoelectrodes. The results obtained in the study suggest that the PFC system involving dual heterojuntion PANI/TiO$_2$ photoanode and CuO/Co$_3$O$_4$ photocathode is very effective for wastewater treatment along with simultaneous electricity generation.

Acknowledgments: This work was supported by the International Science & Technology Cooperation Program of China (Nos. 2013DFG50150 and 2016YFE0126300) and the Innovative and Interdisciplinary Team at HUST (2015ZDTD027). The authors thank the Analytical and Testing Center of HUST for the use of SEM, XRD, TEM, FTIR and DRS equipments.

Author Contributions: Yanrong Zhang and Muthu Murugananthan conceived and designed the experiments; Wenjuan Liao and Na Sun performed the experiments; Honghui Pan contributed analysis tools; Wenjuan Liao wrote the paper.

Conflicts of Interest: The authors declare no conflict of interest.

References

1. Feng, Y.; Lee, H.; Wang, X.; Liu, Y.; He, W. Continuous electricity generation by a graphite granule baffled air–cathode microbial fuel cell. *Bioresour. Technol.* **2010**, *101*, 632–638. [CrossRef] [PubMed]
2. Strataki, N.; Antoniadou, M.; Dracopoulos, V.; Lianos, P. Visible-light photocatalytic hydrogen production from ethanol–water mixtures using a Pt–CdS–TiO$_2$ photocatalyst. *Catal. Today* **2010**, *151*, 53–57. [CrossRef]
3. Liu, Y.; Li, J.; Zhou, B.; Li, X.; Chen, H.; Chen, Q.; Wang, Z.; Li, L.; Wang, J.; Cai, W. Efficient electricity production and simultaneously wastewater treatment via a high-performance photocatalytic fuel cell. *Water Res.* **2011**, *45*, 3991–3998. [CrossRef] [PubMed]
4. Bayati, M.R.; Golestani-Fard, F.; Moshfegh, A.Z. Visible photodecomposition of methylene blue over micro arc oxidized WO$_3$–loaded TiO$_2$ nano-porous layers. *Appl. Catal. A* **2010**, *382*, 322–331. [CrossRef]
5. Lu, B.; Ma, X.; Pan, C.; Zhu, Y. Photocatalytic and photoelectrochemical properties of in situ carbon hybridized BiPO$_4$ films. *Appl. Catal. A* **2012**, *435–436*, 93–98. [CrossRef]
6. Chen, D.; Ye, J. Hierarchical WO$_3$ Hollow Shells: Dendrite, Sphere, Dumbbell, and Their Photocatalytic Properties. *Adv. Funct. Mater.* **2008**, *18*, 1922–1928. [CrossRef]
7. Antoniadou, M.; Kondarides, D.; Labou, D.; Neophytides, S.; Lianos, P. An efficient photoelectrochemical cell functioning in the presence of organic wastes. *Sol. Energy Mater. Sol. Cells* **2010**, *94*, 592–597. [CrossRef]
8. Lianos, P. Production of electricity and hydrogen by photocatalytic degradation of organic wastes in a photoelectrochemical cell: The concept of the Photofuelcell: A review of a re-emerging research field. *J. Hazard. Mater.* **2011**, *185*, 575–590. [CrossRef] [PubMed]
9. Liu, Y.; Li, J.; Zhou, B.; Lv, S.; Li, X.; Chen, H.; Chen, Q.; Cai, W. Photoelectrocatalytic degradation of refractory organic compounds enhanced by a photocatalytic fuel cell. *Appl. Catal. B* **2012**, *111–112*, 485–491. [CrossRef]
10. Liu, Y.; Li, J.; Zhou, B.; Chen, H.; Wang, Z.; Cai, W. A TiO$_2$-nanotube-array-based photocatalytic fuel cell using refractory organic compounds as substrates for electricity generation. *Chem. Commun.* **2011**, *47*, 10314–10316. [CrossRef] [PubMed]
11. Georgieva, J.; Valova, E.; Armyanov, S.; Philippidis, N.; Poulios, I.; Sotiropoulos, S. Bi-component semiconductor oxide photoanodes for the photoelectrocatalytic oxidation of organic solutes and vapours: A short review with emphasis to TiO$_2$–WO$_3$ photoanodes. *J. Hazard. Mater.* **2012**, *211–212*, 30–46. [CrossRef] [PubMed]
12. Li, K.; Zhang, H.; Tang, Y.; Ying, D.; Xu, Y.; Wang, Y.; Jia, J. Photocatalytic degradation and electricity generation in a rotating disk photoelectrochemical cell over hierarchical structured BiOBr film. *Appl. Catal. B* **2015**, *164*, 82–91. [CrossRef]
13. Wang, D.; Li, Y.; Li Puma, G.; Wang, C.; Wang, P.; Zhang, W.; Wang, Q. Dye-sensitized photoelectrochemical cell on plasmonic Ag/AgCl @ chiral TiO$_2$ nanofibers for treatment of urban wastewater effluents, with simultaneous production of hydrogen and electricity. *Appl. Catal. B* **2015**, *168–169*, 25–32. [CrossRef]
14. Du, Y.; Feng, Y.; Qu, Y.; Liu, J.; Ren, N.; Liu, H. Electricity Generation and Pollutant Degradation Using a Novel Biocathode Coupled Photoelectrochemical Cell. *Environ. Sci. Technol.* **2014**, *48*, 7634–7641. [CrossRef] [PubMed]
15. Li, J.; Li, J.; Chen, Q.; Bai, J.; Zhou, B. Converting hazardous organics into clean energy using a solar responsive dual photoelectrode photocatalytic fuel cell. *J. Hazard. Mater.* **2013**, *262*, 304–310. [CrossRef] [PubMed]
16. Chen, Q.; Li, J.; Li, X.; Huang, K.; Zhou, B.; Shangguan, W. Self-Biasing Photoelectrochemical Cell for Spontaneous Overall Water Splitting under Visible-Light Illumination. *ChemSusChem* **2013**, *6*, 1276–1281. [CrossRef] [PubMed]
17. Paracchino, A.; Laporte, V.; Sivula, K.; Grätzel, M.; Thimsen, E. Highly active oxide photocathode for photoelectrochemical water reduction. *Nat. Mater.* **2011**, *10*, 456–461. [CrossRef] [PubMed]
18. Bai, J.; Wang, R.; Li, Y.; Tang, Y.; Zeng, Q.; Xia, L.; Li, X.; Li, J.; Li, C.; Zhou, B. A solar light driven dual photoelectrode photocatalytic fuel cell (PFC) for simultaneous wastewater treatment and electricity generation. *J. Hazard. Mater.* **2016**, *311*, 51–62. [CrossRef] [PubMed]
19. Wang, H.; Zhang, L.; Chen, Z.; Hu, J.; Li, S.; Wang, Z.; Liu, J.; Wang, X. Semiconductor heterojunction photocatalysts: Design, construction, and photocatalytic performances. *Chem. Soc. Rev.* **2014**, *43*, 5234–5244. [CrossRef] [PubMed]

20. Liao, G.; Chen, S.; Quan, X.; Zhang, Y.; Zhao, H. Remarkable improvement of visible light photocatalysis with PANI modified core–shell mesoporous TiO$_2$ microspheres. *Appl. Catal. B* **2011**, *102*, 126–131. [CrossRef]
21. Li, X.; Wang, D.; Cheng, G.; Luo, Q.; An, J.; Wang, Y. Preparation of polyaniline-modified TiO$_2$ nanoparticles and their photocatalytic activity under visible light illumination. *Appl. Catal. B* **2008**, *81*, 267–273. [CrossRef]
22. Liu, J.; Wang, D.; Wang, M.; Kong, D.; Zhang, Y.; Chen, J.F.; Dai, L. Uniform Two-dimensional Co$_3$O$_4$ Porous Sheets: Facile Synthesis and Enhanced Photocatalytic Performance. *Chem. Eng. Technol.* **2016**, *39*, 891–898. [CrossRef]
23. Yehezkeli, O.; de Oliveira, D.R.; Cha, J.N. Electrostatically Assembled CdS–Co$_3$O$_4$ Nanostructures for Photo-assisted Water Oxidation and Photocatalytic Reduction of Dye Molecules. *Small* **2015**, *11*, 668–674. [CrossRef] [PubMed]
24. Shaabani, B.; Alizadeh-Gheshlaghi, E.; Azizian-Kalandaragh, Y.; Khodayari, A. Preparation of CuO nanopowders and their catalytic activity in photodegradation of Rhodamine-B. *Adv. Powder Technol.* **2014**, *25*, 1043–1052. [CrossRef]
25. Shi, W.; Chopra, N. Surfactant-free synthesis of novel copper oxide (CuO) nanowire–cobalt oxide (Co$_3$O$_4$) nanoparticle heterostructures and their morphological control. *J. Nanopart. Res.* **2011**, *13*, 851–868. [CrossRef]
26. Shi, W.; Chopra, N. Controlled fabrication of photoactive copper oxide–cobalt oxide nanowire heterostructures for efficient phenol photodegradation. *ACS Appl. Mater. Interfaces* **2012**, *4*, 5590–5607. [CrossRef] [PubMed]
27. Yavuz, A.G.; Gök, A. Preparation of TiO$_2$/PANI composites in the presence of surfactants and investigation of electrical properties. *Synth. Met.* **2007**, *157*, 235–242. [CrossRef]
28. Liao, W.; Zhang, Y.; Zhang, M.; Murugananthan, M.; Yoshihara, S. Photoelectrocatalytic degradation of microcystin-LR using Ag/AgCl/TiO$_2$ nanotube arrays electrode under visible light irradiation. *Chem. Eng. J.* **2013**, *231*, 455–463. [CrossRef]
29. Chen, S.; Hu, Y.; Meng, S.; Fu, X. Study on the separation mechanisms of photogenerated electrons and holes for composite photocatalysts g-C$_3$N$_4$-WO$_3$. *Appl. Catal. B* **2014**, *150–151*, 564–573. [CrossRef]
30. Yang, Y.; Guo, W.; Guo, Y.; Zhao, Y.; Yuan, X.; Guo, Y. Fabrication of Z-scheme plasmonic photocatalyst Ag@ AgBr/g-C$_3$N$_4$ with enhanced visible-light photocatalytic activity. *J. Hazard. Mater.* **2014**, *271*, 150–159. [CrossRef] [PubMed]
31. Yang, J.; Liao, W.; Liu, Y.; Murugananthan, M.; Zhang, Y. Degradation of Rhodamine B using a Visible-light driven Photocatalytic Fuel Cell. *Electrochim. Acta* **2014**, *144*, 7–15. [CrossRef]
32. Spadavecchia, F.; Cappelletti, G.; Ardizzone, S.; Ceotto, M.; Falciola, L. Electronic structure of pure and N-doped TiO$_2$ nanocrystals by electrochemical experiments and first principles calculations. *J. Phys. Chem. C* **2011**, *115*, 6381–6391. [CrossRef]
33. Gu, J.; Yan, Y.; Krizan, J.W.; Gibson, Q.D.; Detweiler, Z.M.; Cava, R.J.; Bocarsly, A.B. p-Type CuRhO$_2$ as a self-healing photoelectrode for water reduction under visible light. *J. Am. Chem. Soc.* **2014**, *136*, 830–833. [CrossRef] [PubMed]
34. Zhou, H.; Zhang, Y. Enhanced electrochemical performance of manganese dioxide spheres deposited on a titanium dioxide nanotube arrays substrate. *J. Power Sources* **2014**, *272*, 866–879. [CrossRef]
35. Huang, X.; Cao, T.; Liu, M.; Zhao, G. Synergistic Photoelectrochemical Synthesis of Formate from CO$_2$ on {121} Hierarchical Co$_3$O$_4$. *J. Phys. Chem. C* **2013**, *117*, 26432–26440. [CrossRef]
36. Shen, Q.; Chen, Z.; Huang, X.; Liu, M.; Zhao, G. High-yield and selective photoelectrocatalytic reduction of CO$_2$ to formate by metallic copper decorated Co$_3$O$_4$ nanotube arrays. *Environ. Sci. Technol.* **2015**, *49*, 5828–5835. [CrossRef] [PubMed]
37. Lee, C.Y.; Lee, K.; Schmuki, P. Anodic Formation of Self-Organized Cobalt Oxide Nanoporous Layers. *Angew. Chem. Int. Ed.* **2013**, *52*, 2077–2081. [CrossRef] [PubMed]

© 2018 by the authors. Licensee MDPI, Basel, Switzerland. This article is an open access article distributed under the terms and conditions of the Creative Commons Attribution (CC BY) license (http://creativecommons.org/licenses/by/4.0/).

Low-Temperature Sol-Gel Synthesis of Nitrogen-Doped Anatase/Brookite Biphasic Nanoparticles with High Surface Area and Visible-Light Performance

Liang Jiang, Yizhou Li, Haiyan Yang, Yepeng Yang, Jun Liu, Zhiying Yan, Xiang Long, Jiao He and Jiaqiang Wang *

National Center for International Research on Photoelectric and Energy Materials (MOST), Yunnan Provincial Collaborative Innovation Center of Green Chemistry for Lignite Energy, Yunnan Province Engineering Research Center of Photocatalytic Treatment of Industrial Wastewater, The Universities' Center for Photocatalytic Treatment of Pollutants in Yunnan Province, School of Energy, School of Chemical Sciences & Technology, Yunnan University, Kunming 650091, China; liangjiang_ynu@163.com (L.J.); zh111111ou@sina.com (Y.L.); ashoulu@sina.com (H.Y.); mondaysunday1234@163.com (Y.Y.); 18468068607@163.com (J.L.); zhyyan@ynu.edu.cn (Z.Y.); lxbl1991@163.com (X.L.); hejiao@ynu.edu.cn (J.H.)
* Correspondence: jqwang@ynu.edu.cn; Tel.: +86-871-6503-1567

Received: 21 October 2017; Accepted: 30 November 2017; Published: 4 December 2017

Abstract: Nitrogen doping in combination with the brookite phase or a mixture of TiO_2 polymorphs nanomaterials can enhance photocatalytic activity under visible light. Generally, nitrogen-dopedanatase/brookite mixed phases TiO_2 nanoparticles obtained by hydrothermal or solvothermal method need to be at high temperature and with long time heating treatment. Furthermore, the surface areas of them are low (<125 m^2/g). There is hardly a report on the simple and direct preparation of N-doped anatase/brookite mixed phase TiO_2 nanostructures using sol-gel method at low heating temperature. In this paper, the nitrogen-doped anatase/brookite biphasic nanoparticles with large surface area (240 m^2/g) were successfully prepared using sol-gel method at low temperature (165 °C), and with short heating time (4 h) under autogenous pressure. The obtained sample without subsequent annealing at elevated temperatures showed enhanced photocatalytic efficiency for the degradation of methyl orange (MO) with 4.2-, 9.6-, and 7.5-fold visible light activities compared to P25 and the amorphous samples heated in muffle furnace with air or in tube furnace with a flow of nitrogen at 165 °C, respectively. This result was attributed to the synergistic effects of nitrogen doping, mixed crystalline phases, and high surface area.

Keywords: anatase/brookite biphasic; nitrogen-doping; sol-gel method; visible light photocatalysis; degradation of dyes

1. Introduction

Heterogeneous photocatalytic processes involving TiO_2 semiconductor particles have been shown to be a promising process for the treatment of dye effluents [1]. However, large band gap energy (3.2 eV) for anatase TiO_2 limits its practical application for natural solar applications [2]. To develop more light-efficient catalysts, there is an urgent need to develop photocatalytic systems which are able to operate effectively under visible light irradiation. A number of systems have been reported to improve the visible-light activity of TiO_2. Meanwhile, selecting the reasonable substrate and activity test are helpful to systematically and comprehensively assess the photocatalytic efficiency of the catalysts [3]. Nitrogen-doped (N-doped) TiO_2 is one of the most typical examples of the visible-light photocatalysts, which is due to nitrogen doping can decrease the band gap energy and enhance the photoactivity of TiO$_2$ in the visible spectral range [4,5]. However, the low reactivity and quantum efficiency of N-doped TiO$_2$ limit its practical application [6,7].

On the other hand, TiO$_2$ exists in three main polymorphs, which are anatase, rutile, and brookite [8,9]. Phase mixing is well recognized as the most promising strategy for quantum efficiency improvement, which can be due to the enhanced charge carrier separation [6,10–14]. Particularly, it has been proven that the mixed phase of anatase/rutile TiO$_2$ has synergistic effects and higher photocatalytic activity as compared to pure phase of either in anatase or rutile [15,16]. In contrast to anatase/rutile biphasic nanoparticles which have been intensively studied, the photocatalytic study of brookite and its phase mixing is quite limited, though it has been reported that anatase/brookite mixed-phase TiO$_2$ has higher activity in visible light than P25 [8]. The reason may be mainly due to the difficulties in synthesis [17]. For example, anatase–brookite composite nanocrystals were synthesized by a sonochemical sol-gel method at very high heating temperature (500 °C) [12,18]. Highly crystalline phase-pure brookite and anatase/brookite mixed-phase TiO$_2$ nanostructures were synthesized via a simple hydrothermal method with titanium sulphide as the precursors in sodium hydroxide solutions [19]. Interestingly, anatase-brookite heterojunction TiO$_2$ photocatalysts were purposefully tailored by introducing different glycine concentrations through hydrothermal treatment at 200 °C for 20 h [20].

It is expected that a strategy coupling a binary structure with nitrogen doping could bring enhanced photocatalytic properties of TiO$_2$. Recently, N-doped anatase/rutile TiO$_2$ nanoparticles have been designed and synthesized [19,21]. Anatase/brookite mixed-phase nitrogen-doped TiO$_2$ nanoparticles were also synthesized by a facile solvothermal route [22]. Interestingly, nitrogen plasma treatment was employed to prepare N-doped nanoporous TiO$_2$ with large surface area and high-crystalline anatase/brookite phase [23].

Generally, a semiconductor catalyst with large specific surface area is beneficial for efficient photocatalysis, while in most synthetic processes, TiO$_2$ with the brookite phase or a mixture of TiO$_2$ polymorphs obtained hydrothermally at high temperature and with long time heat treatment have low surface area [14,19,23,24]. Hence, it is challenging to synthesize N-doped anatase/brookite TiO$_2$ photocatalyst with large surface area and enhanced visible light activity at low temperature via simple and direct synthetic method.

Sol-gel is one of the most prominent methods used to prepare mixed phases of anatase/rutile TiO$_2$ nanoparticles due to its simplicity and low equipment requirements. However, there are few reports on the simple and direct preparation of N-doped anatase/brookite mixed phase TiO$_2$ nanostructures using sol-gel method at low heating temperature [15,16]. The goal of the present work is to synthesize anatase/brookite biphasic TiO$_2$ nanoparticles by direct introduction of nitrogen in TiO$_2$ lattice crystal during the sol-gel preparation at low temperature. In this work, the degradation of methyl orange (MO) in aqueous solution under visible light irradiations was selected to test the enhanced photocatalytic efficiency. It has been reported that amorphous TiO$_2$ or a mixture composed of crystalline and amorphous TiO$_2$ has high activity for the photocatalytic degradation of pollutants [25,26]. However, synthesized nitrogen-doped anatase/brookite biphasic nanoparticles of this work exhibited much higher photocatalytic efficiency than the prepared amorphous samples.

2. Results and Discussion

2.1. Syntheses and Characterizations

The synthesis process of this work was modified from a typical sol-gel method by using HNO$_3$-catalyzed hydrolysis step of titanium tetraisopropoxide (TTIP) to reduce the hydrolysis rates [27]. Generally, heating is required to prepare crystalline TiO$_2$. If low heating temperature in the range of 180–200 °C was employed in hydrothermal or solvothermal method, longer time (3–48 h) would be needed. Nevertheless, the obtained TiO$_2$ samples still have low surface area (<125 m^2/g) [18,21,23]. Moreover, a supercritical drying process was often used in the conventional sol-gel method [28]. By contrast, herein the aged gels were heated under nitrogen atmosphere with a much lower final autogenous pressure (about 350 psi), heating temperature (165 °C), and shorter heating time (4 h).

The crystal structures of sample NA-185 and NA-165 with anatase and brookite phases were identified by X-ray diffraction (XRD), as shown in Figure 1. The diffraction peaks of 2θ values at 25.3°, 37.8°, 48.1°, 54.9°, 62.8°, 69.8°, 75.4°, 82.8° are assigned to the (101), (004), (200), (204), (220), (215), and (224) planes of anatase TiO_2 (JCPDS 21-1272). Due to the overlapping of the planes of brookite (120), brookite (111), and anatase (101), the existence of the brookite phase was determined by the brookite (121) plane at 30.8° (JCPDS 29-1360). Both NA-145 and HA-165 are phase-pure anatase. From the XRD peak intensities [29], the brookite phase contents of sample NA-185 and NA-165 were calculated to be ~10% and ~6%, respectively. The crystal size was calculated by Scherrer equation (Table 1).

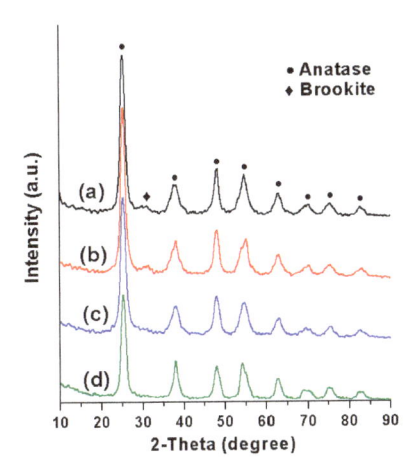

Figure 1. X-ray diffraction (XRD) patterns of as-prepared samples: (**a**) NA-185, (**b**) NA-165, (**c**) NA-145, (**d**) HA-165.

Table 1. The characteristics and the apparent first-order rate constant K (min^{-1}) of samples.

Samples	Anatase		Brookite		S_{BET} (m^2/g)	K (min^{-1})
	Crystal Size [a] (nm)	Content [b] (%)	Crystal Size [a] (nm)	Content [b] (%)		
NA-185	6.7	90	7.3	10	239	0.023
NA-165	7.5	94	7.2	6	240	0.021
NA-145	6.5	100	-	-	249	0.015
HA-165	7.9	100	-	-	216	0.006
TF-165	-	-	-	-	443	0.002
MF-165	-	-	-	-	407	0.003
P25	-	-	-	-	50	0.005

[a] Determined by the Scherrer equation; [b] Calculated using the formula in reference [29].

The morphology and particle size of the samples revealed by Scanning electron microscopy (SEM) and Transmission electron microscopy (TEM) analysis (Figures S1 and S2). NA-185, NA-165, NA-145, and HA-165 all show aggregates consisting of small spheroidal nanoparticles with average size of approximately 6–8 nm, which was in agreement with the results calculated by Scherrer equation. To further confirm the existence of anatase and brookite phases, high-resolution TEM (HRTEM) analysis of NA-165 was performed. As shown in Figure 2a,b, the lattice fringes of 0.35 nm and 0.29 nm match the anatase (101) and brookite (121) plane, respectively. The results are in agreement with the XRD observations.

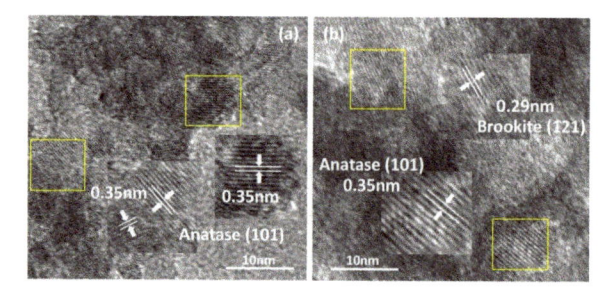

Figure 2. High-resolution TEM (HRTEM) images of NA-165. The lattice fringes of 0.35 nm and 0.29 nm match (**a**) the anatase (101) and (**b**) brookite (121) plane.

The X-ray photoelectron spectroscopy (XPS) measurements reveal the surface compositions and chemical states of the samples with the presence of N, O, Ti, and C. The N 1s peak of NA-165 at 400.1 eV can be attributed to the interstitial nitrogen in the form of Ti-O-N or Ti-N-O bonds (Figure 3a) [30]. Moreover, the Ti 2p2/3 and Ti 2p1/2 core levels were located at 458.4 and 464.2 eV (Figure 3b), which shift toward lower binding energies as compared to the reported pure TiO_2 due to the nitrogen doping [31]. The nitrogen doping percentages of NA-185, NA-165, NA-145, and HA-165 were 0.66, 0.63, 0.73, and 0.52 at.%, respectively. Since the nitrogen content of NA-165 prepared under N_2 using HNO_3 as catalyst is higher than that of HA-165 prepared under N_2 via the similar method under N_2 but using HCl instead of HNO_3 as catalyst, it implied that the nitrogen source in NA-165 may be from both N_2 and HNO_3 [22,32]. The O 1s XPS spectra of NA-165 shown in Figure 3c displays two peaks at 530.2 and 531.8 eV, which was attributed to the Ti–O bond and Ti-O-N or Ti-N-O, respectively [33]. The XPS results along with XRD patterns and HRTEM images reveal that nitrogen-doped anatase/brookite biphasic nanoparticles were successfully synthesized.

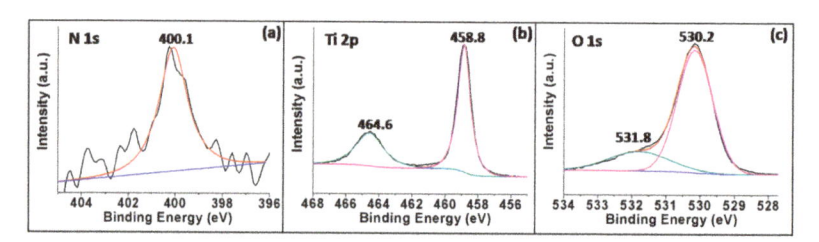

Figure 3. X-ray photoelectron spectroscopy (XPS) spectra of (**a**) N 1s, (**b**) Ti 2p, (**c**) O 1s region for NA-165.

The nitrogen adsorption-desorption isotherms shown in Figure S3 are all Type IV, implying that the samples may have mesoporous structures. The surface areas, average pore size, and pore volumes of the samples are summarized in Table 1 and Table S1. Obviously, the surface areas of the biphasic samples changed little with the increase of heating temperature, since the surface areas of NA-145, NA-165, and NA-185 were 249, 240, and 239 m^2/g, respectively. HA-165 prepared using HCl instead of HNO_3 as catalyst had lower surface area (216 m^2/g). Compared with other methods, the employed heating temperature of this work was much lower, and the heat treatment time was shorter. Nevertheless, the surface area of NA-165 was also much higher than those of many other types of nitrogen-doped anatase/brookite biphasic TiO_2 except for the one treated with nitrogen plasma [22].

The UV-vis diffuse reflectance spectra of NA-185, NA-165, NA-145, and HA-165 are shown in Figure 4 using P25 as a control group. The absorbance of the N-doped samples was stronger than that of P25 in the visible light region. The band gap energies of NA-185, NA-165, NA-145, HA-165, and P25 were 3.05, 3.03, 3.01, 3.09, and 3.12 eV, respectively, which were calculated from Equation (1):

$$Eg = 1240/\lambda, \tag{1}$$

where Eg and λ are the band gap energy (eV) and wavelength of adsorption edge (nm), respectively. The narrower band gap and stronger visible-light response of the samples can be ascribed to the effect of the nitrogen doping [3,31]. Among the two biphasic samples, the band gap energies were increased with increasing brookite content. The reason may be due to the band gap of brookite is larger than anatase [18].

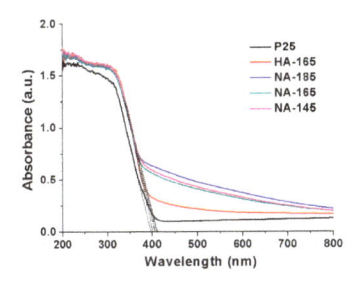

Figure 4. The UV-vis diffuse reflectance spectra of NA-145, NA-165, NA-185, HA-165, and P25.

2.2. Photocatalytic Activity

The visible light photocatalytic activities of as-prepared samples were tested by photodegradation of MO (10 mg/L). For comparison, we have also studied the photocatalytic activities of P25 and the two samples prepared with the similar sol-gel method but heated at 165 °C in a muffle furnace with air (MF-165) or a resistance-heated tube furnace with a flow of nitrogen (TF-165). Figure 5a shows the removal rates of MO for NA-185, NA-165, NA-145, HA-165, P25, MF-165, and TF-165 are 95%, 92%, 83%, 53%, 49%, 27%, and 31%, respectively. The dark reaction adsorption rates of samples are all less than 8%, which implied that the removal of MO is mainly attributed to photocatalytic degradation rather than adsorption. Figure S4 shows the nitrogen-doped anatase/brookite biphasic samples of NA-185 (94%) and NA-165 (91%) with similar photocatalytic degradation rate of MO, which are higher than the other samples. The apparent first-order rate constant K (min^{-1}) for NA-165 (0.021) is close to that of NA-185 (0.023), which is about 1.4, 3.5, 4.2, 9.6, and 7.5 times higher than those of NA-145, HA-165, P25, MF-165, and TF-165, respectively (Figure 5b and Table 1). Thus, 165 °C was chosen as a reasonable heating treatment temperature.

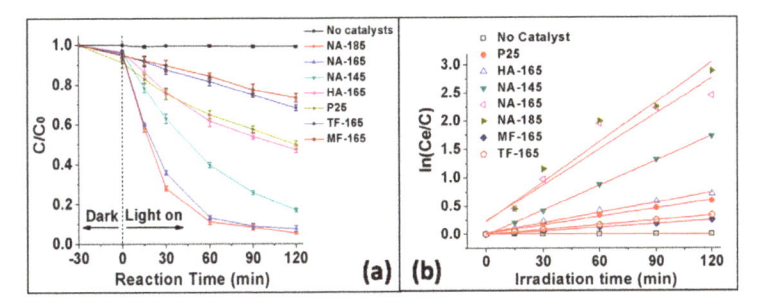

Figure 5. (a) Removal curves of methyl orange (MO). Error bars represent the standard deviation from three measurements; (b) Apparent first-order kinetics plot for the photocatalytic degradation of MO over different samples.

Table 2 summarizes the preparation methods, surface area, and visible-light photocatalytic activity of nitrogen-doped anatase/brookite biphasic TiO$_2$ reported in recent years. The 4.2-fold visible light activity enhancement as compared to P25 suggests that NA-165 is a potential highly efficient photocatalyst. By contrast, if the aged gel was heated in amuffle furnace with air or obtained in a resistance-heated tube furnace with a flow of nitrogen at the same temperature (165 °C), respectively, only amorphous samples were obtained in spite of large surface area (Figure S5 and Table 1). Meanwhile, they were much less active compared to NA-165. This implies that the crystallinity may play a more important role. Moreover, the presence of brookite in the mixture can reduce the recombination of hole–electron pairs. The band gap was also widened with increasing brookite content [34]. This is why NA-165 (3.03 eV, 6% brookite content) and NA-185 (3.05 eV, 10% brookite content) exhibited similar MO photocatalytic degradation activity.

The photocatalytic stability of NA-165 was tested by cycling experiments. For each cycling run, NA-165 was separated by centrifugation, and dried at 90 °C. As shown in Figure 6, there was no significant decrease of photocatalytic degradation rate after three cycling runs. This result suggested that NA-165 was a stable photocatalyst for organic dye degradation under visible light.

Table 2. Comparison of nitrogen-doped anatase/brookite biphasic TiO$_2$ prepared by various methods.

Sample	Surface Area (m^2/g)	Preparation Method	Application	The Times of K (min^{-1}) to P25	Reference
Nitrogen-doped TiO$_2$ nanorods with anatase/brookite structures	51.1	Hydrothermal synthesis 200 °C, 48 h	Degradation of MO and 4-chlorophenol (4-CP)	2.3, 2.7	[19]
Nitrogen-doped anatase/brookite titania	124.4	Solvothermal synthesis 190 °C, 3 h	Degradation of MO		[24]
Anatase–brookite mixed-phase N-doped TiO$_2$ nanoparticles	76.2	Solvothermal synthesis 180 °C, 48 h	Degradation of Methylene blue (MB)		[22]
Nitrogen-doped TiO$_2$ film		Solvothermal synthesis 180 °C, 18 h	Degradation of MB		[35]
Bicrystalline (anatase/brookite) nanoporous nitrogen-doped TiO$_2$	375.9	Plasma treatment 0.5 h	Degradation of Rhodamine B (RhB)		[23]
Nitrogen-doped anatase/brookite biphasic nanoparticles	240	Sol-gel synthesis 165 °C, 4 h	Degradation of MO	4.2	This work

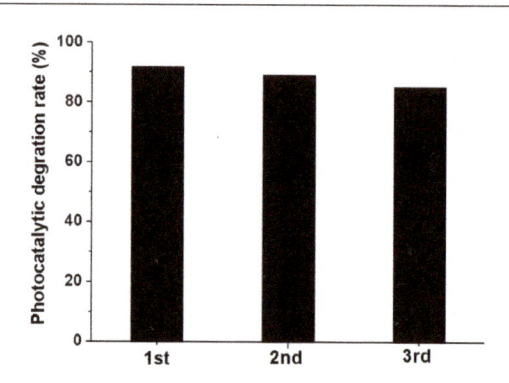

Figure 6. Cycling run in the photocatalytic degradation of MO under visible light over NA-165.

2.3. Possible Reasons for the Enhancement of the Visible-Light Performance

It is interesting to evoke some reasons why NA-165 has high visible-light performance, though mechanism of the enhancement is still far from understood. The first explanation is that the absorption edge of NA-165 shifts to the visible-light range, and then they possess narrower band gap, and have definite absorptions in the visible region due to the presence of nitrogen-doping, which has been confirmed by UV-vis diffuse reflectance spectra and XPS study. Secondly, the anatase/brookite biphasic nanoparticles are aggregated closely, as shown in Figures S1 and S2. The intimate contact can facilitate inter particle charge transfer from brookite to anatase and reduce the recombination of electron–hole pairs. Thirdly, the large surface area can provide more active sites and improve the diffusion and migration of MO in the process of photodegradation [36]. Furthermore, the competitive diffusion of the H_2O and dye molecules, dye molecule structure, and photocatalytic degradation route are also the factors influencingthe photocatalytic process [37]. The photodegradation of MO under visible light was mainly driven by the active species $O_2^{\bullet-}$, h^+, and $^{\bullet}OH$ [38].

3. Materials and Methods

3.1. Synthesis

Titanium tetraisopropoxide (TTIP, $\geq$97%, Sigma-Aldrich, St. Louis, MO, USA) was of chemical grade. Acetone ($\geq$99.5%, Tianjin Fengchuan Chemical Reagent Technologies Co., Ltd., Tianjin, China), HNO_3 (65%, Xilong Scientific Co., Ltd., Shantou, China), and acetylacetone ($\geq$98%, Tianjin Fengchuan Chemical Reagent Technologies Co., Ltd., Tianjin, China) were of analytical grade. All chemicals were used without further purification.

Nitrogen-doped anatase/brookite biphasic nanoparticles were prepared with a sol-gel process modified from a sol-gel combined solvothermal route [24]. Titanium tetraisopropoxide, acetone, HNO_3 and acetylacetone with the volume ratio of 6.5:20:0.11:0.54 were mixed in a glass beaker. A mixture solution of deionized water and acetone (volume ratio of 1.2:7.5) was then added dropwise with vigorous stirring until reaching the gelling point. The gels were placed into a quartz-lined stainless-steel autoclave after being aged for 24 h at room temperature. Then, the temperature of the autoclave was increased to and held at 145, 165, or 185 °C for 4 h under nitrogen atmosphere after flushing the autoclave with nitrogen gas. The initial and final pressures were under atmospheric and autogenous pressure, respectively. After the heat treatment ended, the pressure was released quickly to remove the solvent vapour. The obtained material was cooled down to room temperature by nitrogen purging before being washed with deionized water and dried in vacuum at 90 °C for 4 h. According to the heat treatment temperature, the samples were denoted as NA-145, NA-165, and NA-185, respectively. If HNO_3 was replaced by the same volume of HCl in the process of preparation, then the sample was denoted as HA-165.

3.2. Photocatalytic Activity

In each experiment, 50 mg photocatalysts and 50 mL of MO solution (10 mg/L) was placed in a glass vessel with a cooling water jacket and quartz cover. The suspensions were stirred in the dark for 30 min to reach the adsorption–desorption equilibrium. Then, the system was exposed for 120 min under visible light irradiation provided by a 500 W Xe lamp with a 420 nm cut off filter. At certain time intervals, 3 mL of the suspensions was collected and centrifuged (10,000 rpm, 20 min) to remove the photocatalysts. The separated solution was analysed and the maximum absorption was recorded at 464 nm by a spectrophotometer (Shimadzu UV-2600, Kyoto, Japan).

The removal rate of MO was calculated using Equation (2):

$$\text{removal rate} = C/C_0, \qquad (2)$$

where C and C_0 are the initial and instantaneous absorbance of MO at 464 nm.

The photocatalytic degradation rates of MO and the first-order rate constant K (min^{-1}) were calculated, respectively, using the Equations (3) and (4):

$$\text{photocatalytic degradation rate} = C/C_e, \qquad (3)$$

$$\ln(C_e/C) = Kt, \qquad (4)$$

where Ce and C are the adsorption–desorption equilibrium absorbance and instantaneous absorbance of MO at 464 nm, respectively. t is the irradiation time.

3.3. Characterizations

X-ray powder diffraction (XRD, Rigaku Co., Tokyo, Japan) analysis was conducted on a D/max-3B spectrometer with Cu Kα radiation at a range from 10° to 90° (2θ). Brunauer–Emmett–Teller (BET) surface area, pore volume, and pore size were measured by nitrogen adsorption/desorption using a Micromeritics Tristar II Surface Area and Porosity Analyzer (Micromeritics, Norcross, GA, USA). Transmission electron microscopy (TEM) was conducted on a Hitachi H-800 instrument (Japan Electron Optics Laboratory Co., Ltd., Tokyo, Japan). Scanning electron microscopy (SEM) images were taken on a FEIQuanta200FEG microscope (FEI, Hillsboro, OR, USA). X-ray photoelectron spectroscopy (XPS) was recorded using a Thermo Fisher Scientific K-Alpha$^+$ XPS system with Al Kα radiation and adventitious C1s peak (284.8 eV) calibration (Thermo Fisher Scientific Inc., Waltham, MA, USA). UV-Vis diffuse reflectance spectra were measured on a UV-2600 photometer (Shimadzu Corp., Kyoto, Japan).

4. Conclusions

The nitrogen-doped anatase/brookite biphasic nanoparticles with large surface area (240 m^2/g) were successfully prepared during the sol-gel preparation at low temperature (165 °C, 4 h). The sample obtained without subsequent annealing at elevated temperature, which exhibited enhanced visible-light photocatalytic efficiency for the degradation of MO with 4.2-, 9.6-, and 7.5-fold visible light activities as compared to P25, MF-165, and TF-165, respectively. This was attributed to nitrogen doping, mixed crystalline phase, and high surface area. The recycling experiments suggested that NA-165 was a stable visible-light photocatalyst. The sample and low-temperature synthetic method developed in this work may provide a new pathway to prepare the stable photocatalyst for the degradation of organic dyes under visible light.

Supplementary Materials: The following are available online at www.mdpi.com/2073-4344/7/12/376/s1, Figure S1: Scanning electron microscopy (SEM) images of NA-185 (a), NA-165 (b), NA-145 (c) and HA-145 (d); Figure S2: Transmission electron microscopy (TEM) Transmission electron microscopy (TEM); Figure S3: Nitrogen adsorption–desorption isotherms for the prepared samples; Figure S4: Photocatalytic degradation of MO under visible light over different samples; Figure S5: XRD patterns of MF-165 and TF-165; Table S1: The Brunauer–Emmett–Teller (BET)analysis data of samples.

Acknowledgments: The work was supported by National Natural Science Foundation of China (Project 21403190, 21573193, 21367024, 21464016 and 21263027). The authors also thank Program for Innovation Team of Yunnan Province and Innovative Research Team (in Science and Technology) in the Universities of Yunnan Province, Key Laboratory of Advanced Materials for Wastewater Treatment of Kunming, the Key project from the Yunnan Educational Committee (Project ZD2012003), Yunnan Provincial Natural Science Foundation (Project 2015FB106) and Yunnan Applied Basic Research Projects (Project 2016FA002) for financial support.

Author Contributions: Jiaqiang Wang and Liang Jiang conceived and designed the experiments; Liang Jiang, Yizhou Li, and Haiyan Yang performed the experiments; Yepeng Yang, Xiang Long and Jun Liu analyzed the data; Liang Jiang, Zhiying Yan, and Jiao He wrote the paper; Jiaqiang Wang and Zhiying Yan modified the paper.

Conflicts of Interest: The authors declare no conflict of interest.

References

1. Zhang, W.; Zou, L.; Wang, L. Photocatalytic TiO$_2$/adsorbent nanocomposites prepared via wet chemical impregnation for wastewater treatment: A review. *Appl. Catal. A* **2009**, *371*, 1–9. [CrossRef]
2. Zhang, Z.; Wang, X.; Long, J.; Gu, Q.; Ding, Z.; Fu, X. Nitrogen-doped titanium dioxide visible light photocatalyst: Spectroscopic identification of photoactive centers. *J. Catal.* **2010**, *276*, 201–214. [CrossRef]
3. Ryu, J.; Choi, W. Substrate-specific photocatalytic activities of TiO$_2$ and multiactivity test for water treatment application. *Environ. Sci. Technol.* **2008**, *42*, 294–300. [CrossRef] [PubMed]
4. Asahi, R.; Morikawa, T.; Ohwaki, T.; Aoki, K.; Taga, Y. Visible-light photocatalysis in nitrogen-doped titanium oxides. *Science* **2001**, *293*, 269–271. [CrossRef] [PubMed]

5. Valentin, C.D.; Finazzi, E.; Pacchioni, G.; Selloni, A.; Livraghi, S.; Paganini, M.C.; Giamello, E. N-doped TiO$_2$: Theory and experiment. *Chem. Phys.* **2007**, *339*, 44–56. [CrossRef]
6. Wang, H.; Gao, X.; Duan, G.; Yang, X.; Liu, X. Facile preparation of anatase–brookite–rutile mixed-phase N-doped TiO$_2$ with high visible-light photocatalytic activity. *J. Environ. Chem. Eng.* **2015**, *3*, 603–608. [CrossRef]
7. Li, H.; Li, J.; Huo, Y. Highly Active TiO$_2$N Photocatalysts Prepared by Treating TiO$_2$ Precursors in NH$_3$/Ethanol Fluid under Supercritical Conditions. *J. Phys. Chem. B* **2006**, *110*, 1559–1565. [CrossRef] [PubMed]
8. Paola, A.D.; Bellardita, M.; Palmisano, L. Brookite, the Least Known TiO$_2$ Photocatalyst. *Catalysts* **2013**, *3*, 36–73. [CrossRef]
9. Reyes-Coronado, D.; Rodríguez-Gattorno, G.; Espinosa-Pesqueira, M.E.; Cab, C.; de Coss, R.; Oskam, G. Phase-pure TiO$_2$ nanoparticles: Anatase, brookite and rutile. *Nanotechnology* **2008**, *19*, 145605. [CrossRef] [PubMed]
10. Yu, J.; Yu, J.C.; Leung, M.K.P.; Ho, W.; Cheng, B.; Zhao, X.; Zhao, J. Effects of acidic and basic hydrolysis catalysts on the photocatalytic activity and microstructures of bimodal mesoporous titania. *J. Catal.* **2003**, *217*, 69–78. [CrossRef]
11. Zhao, B.; Chen, F.; Jiao, Y.; Yang, H.; Zhang, J. Ag^0-loaded brookite/anatase composite with enhanced photocatalytic performance towards the degradation of methyl orange. *J. Mol. Catal. A Chem.* **2011**, *348*, 114–119. [CrossRef]
12. Ozawa, T.; Iwasaki, M.; Tada, H.; Akita, T.; Tanaka, K.; Ito, S. Low-temperature synthesis of anatase–brookite composite nanocrystals: The junction effect on photocatalytic activity. *J. Colloid Interface Sci.* **2005**, *281*, 510–513. [CrossRef] [PubMed]
13. Kandiel, T.A.; Feldhoff, A.; Robben, L.; Dillert, R.; Bahnemann, D.W. Tailored Titanium Dioxide Nanomaterials: Anatase Nanoparticles and Brookite Nanorods as Highly Active Photocatalysts. *Chem. Mater.* **2010**, *22*, 2050–2060. [CrossRef]
14. Mutuma, B.K.; Shao, G.N.; Kim, W.D.; Kim, H.T. Sol–gel synthesis of mesoporous anatase–brookite and anatase–brookite–rutile TiO$_2$ nanoparticles and their photocatalytic properties. *J. Colloid Interface Sci.* **2015**, *442*, 1–7. [CrossRef] [PubMed]
15. Scanlon, D.O.; Dunnill, C.W.; Buckeridge, J.; Shevlin, S.A.; Logsdail, A.J.; Woodley, S.M.; Catlow, C.R.A.; Powell, M.J.; Palgrave, R.G.; Parkin, I.P.; et al. Band alignment of rutile and anatase TiO$_2$. *Nat. Mater.* **2013**, *12*, 798–801. [CrossRef] [PubMed]
16. Mohamed, M.A.; Salleh, W.N.W.; Jaafar, J.; Ismail, A.F.; Nor, N.A.M. Photodegradation of phenol by N-Doped TiO$_2$ anatase/rutile nanorods assembled microsphere under UV and visible light irradiation. *Mater. Chem. Phys.* **2015**, *162*, 113–123. [CrossRef]
17. Nie, X.; Zhuo, S.; Maeng, G.; Sohlberg, K. Doping of TiO$_2$ Polymorphs for Altered Optical and Photocatalytic Properties. *Int. J. Photoenergy* **2009**, *2009*, 294042. [CrossRef]
18. Yu, J.C.; Yu, J.; Ho, W.; Zhang, L. Preparation of highly photocatalytic active nano-sized TiO$_2$ particles via ultrasonic irradiation. *Chem. Commun.* **2001**, *19*, 1942–1943. [CrossRef]
19. Gai, L.; Duan, X.; Jiang, H.; Mei, Q.; Zhou, G.; Tian, Y.; Liu, H. One-pot synthesis of nitrogen-doped TiO$_2$ nanorods with anatase/brookite structures and enhanced photocatalytic activity. *CrystEngComm* **2012**, *14*, 7662–7671. [CrossRef]
20. El-Sheikh, S.M.; Khedr, T.M.; Zhang, G.; Vogiazi, V.; Ismail, A.A.; O'Shea, K.; Dionysiou, D.D. Tailored synthesis of anatase–brookite heterojunction photocatalysts for degradation of cylindrospermopsin under UV–Vis light. *Chem. Eng. J.* **2017**, *310*, 428–436. [CrossRef]
21. Etacheri, V.; Seery, M.K.; Hinder, S.J.; Pillai, S.C. Highly Visible Light Active TiO$_{2-x}$N$_x$ Heterojunction Photocatalysts. *Chem. Mater.* **2010**, *22*, 3843–3853. [CrossRef]
22. Li, L.; Liu, C. Facile Synthesis of Anatase–Brookite Mixed-Phase N-Doped TiO$_2$ Nanoparticles with High Visible-Light Photocatalytic Activity. *Eur. J. Inorg. Chem.* **2009**, *25*, 3727–3733. [CrossRef]
23. Lee, H.U.; Lee, Y.; Lee, S.C.; Park, S.Y.; Son, B.; Lee, J.W.; Lim, C.; Choi, C.; Choi, M.; Lee, S.Y.; et al. Visible-light-responsive bicrystalline (anatase/brookite) nanoporous nitrogen-doped TiO$_2$ photocatalysts by plasma treatment. *Chem. Eng. J.* **2014**, *254*, 268–275. [CrossRef]
24. Liu, J.; Qin, W.; Zuo, S.; Yu, Y.; Hao, Z. Solvothermal-induced phase transition and visible photocatalytic activity of nitrogen-doped titania. *J. Hazard. Mater.* **2009**, *163*, 273–278. [CrossRef] [PubMed]

25. Lin, L.; Huang, J.; Li, X.; Abass, M.A.; Zhang, S. Effective surface disorder engineering of metal oxide nanocrystals forimproved photocatalysis. *Appl. Catal. B* **2017**, *203*, 615–624. [CrossRef]
26. Zimbone, M.; Cacciato, G.; Boutinguiza, M.; Privitera, V.; Grimaldi, M.G. Laser irradiation in water for the novel, scalable synthesis of black TiO_x photocatalyst for environmental remediation. *Beilstein J. Nanotechnol.* **2017**, *8*, 196–202. [CrossRef] [PubMed]
27. Zhao, Z.; Jiao, X.; Chen, D. Preparation of TiO_2 aerogels by a sol-gel combined solvothermal route. *J. Mater. Chem.* **2009**, *19*, 3078–3083. [CrossRef]
28. Wang, J.; Uma, S.; Klabunde, K.J. Visible light photocatalysis in transition metal incorporated titania-silica aerogels. *Appl. Catal. B* **2004**, *48*, 151–154. [CrossRef]
29. Li, J.; Ishigaki, T.; Sun, X. Anatase, Brookite, and Rutile Nanocrystals via Redox Reactions under Mild Hydrothermal Conditions: Phase-Selective Synthesis and Physicochemical Properties. *J. Phys. Chem. C* **2007**, *111*, 4969–4976. [CrossRef]
30. Jagadale, T.C.; Takale, S.P.; Sonawane, R.S.; Joshi, H.M.; Patil, S.I.; Kale, B.B.; Ogale, S.B. N-Doped TiO_2 Nanoparticle Based Visible Light Photocatalyst by Modified Peroxide Sol−Gel Method. *J. Phys. Chem. C* **2008**, *112*, 14595–14602. [CrossRef]
31. Li, J.; Xu, X.; Liub, X.; Qin, W.; Pan, L. Novel cake-like N-doped anatase/rutile mixed phase TiO_2 derived from metal-organic frameworks for visible light photocatalysis. *Ceram. Int.* **2017**, *43*, 835–840. [CrossRef]
32. Zhang, Y.C.; Yang, M.; Zhang, G.; Dionysiou, D.D. HNO_3-involved one-step low temperature solvothermal synthesis of N-doped TiO_2 nanocrystals for efficient photocatalytic reduction of Cr(VI) in water. *Appl. Catal. B* **2013**, *142*, 249–258. [CrossRef]
33. György, E.; del Pino, A.P.; Serra, P.; Morenza, J.L. Depth profiling characterisation of the surface layer obtained by pulsed Nd:YAG laser irradiation of titanium in nitrogen. *Surf. Coat. Technol.* **2003**, *173*, 265–270. [CrossRef]
34. Zhao, H.; Liu, L.; Andinobc, J.M.; Li, Y. Bicrystalline TiO_2 with controllable anatase–brookite phase content for enhanced CO_2 photoreduction to fuels. *J. Mater. Chem. A* **2013**, *1*, 8209–8216. [CrossRef]
35. Wang, J.; Mao, B.; Gole, J.L.; Burda, C. Visible-light-driven reversible and switchable hydrophobic to hydrophilic nitrogen-doped titania surfaces: Correlation with photocatalysis. *Nanoscale* **2010**, *2*, 2257–2261. [CrossRef] [PubMed]
36. Yan, Z.; He, J.; Guo, L.; Li, Y.; Duan, D.; Chen, Y.; Li, J.; Yuan, F.; Wang, J. Biotemplated Mesoporous TiO_2/SiO_2 Composite Derived from Aquatic Plant Leaves for Efficient Dye Degradation. *Catalysts* **2017**, *7*, 82. [CrossRef]
37. Miao, Y.; Zhai, Z.; Jiang, L.; Shi, Y.; Yan, Z.; Duan, D.; Zhen, K.; Wang, J. Facile and new synthesis of cobalt doped mesoporous TiO_2 with high visible-light performance. *Powder Technol.* **2014**, *266*, 365–371. [CrossRef]
38. Li, W.; Li, D.; Lin, Y.; Wang, P.; Chen, W.; Fu, X.; Shao, Y. Evidence for the Active Species Involved in the Photodegradation Process of Methyl Orange on TiO_2. *J. Phys. Chem. C* **2012**, *116*, 3552–3560. [CrossRef]

© 2017 by the authors. Licensee MDPI, Basel, Switzerland. This article is an open access article distributed under the terms and conditions of the Creative Commons Attribution (CC BY) license (http://creativecommons.org/licenses/by/4.0/).

catalysts

Article

Effective Photocatalytic Activity of Mixed Ni/Fe-Base Metal-Organic Framework under a Compact Fluorescent Daylight Lamp

Vinh Huu Nguyen [1], Trinh Duy Nguyen [1,*], Long Giang Bach [1], Thai Hoang [2], Quynh Thi Phuong Bui [3], Lam Dai Tran [2,4], Chuong V. Nguyen [5], Dai-Viet N. Vo [6] and Sy Trung Do [7,*]

[1] NTT Hi-tech Institute, Nguyen Tat Thanh University, 300A Nguyen Tat Thanh, District 4, Ho Chi Minh City 755414, Vietnam; nguyenhuuvinh3110@gmail.com (V.H.N.); blgiangntt@gmail.com (L.G.B.)
[2] Institute for Tropical Technology, Vietnam Academy of Science and Technology, 18 Hoang Quoc Viet, Cau Giay, Hanoi 10072, Vietnam; hoangth@itt.vast.vn (T.H.); trandailam@gmail.com (L.D.T.)
[3] Faculty of Chemical Technology, Ho Chi Minh City University of Food Industry, 140 Le Trong Tan, Tan Phu District, Ho Chi Minh City 705800, Vietnam; phuongquynh102008@gmail.com
[4] Graduate University of Science and Technology, Vietnam Academy of Science and Technology, 18 Hoang Quoc Viet, Cau Giay, Hanoi 10072, Vietnam
[5] Department of Materials Science and Engineering, Le Quy Don Technical University, Hanoi 100000, Vietnam; chuongnguyen11@gmail.com
[6] Faculty of Chemical & Natural Resources Engineering, University Malaysia Pahang, Lebuhraya Tun Razak, Gambang 26300, Malaysia; vietvo@ump.edu.my
[7] Institute of Chemistry, Vietnam Academy of Science and Techology, 18 Hoang Quoc Viet, Cau Giay District, Hanoi 10072, Vietnam
* Correspondence: ndtrinh@ntt.edu.vn (T.D.N.); dosyvhh@gmail.com (S.T.D.); Tel.: +84-971-275-356 (T.D.N.); +84-969-937-586 (S.T.D.)

Received: 19 September 2018; Accepted: 15 October 2018; Published: 23 October 2018

Abstract: Mixed Ni/Fe-base metal-organic framework (Ni/Fe-MOF) with different molar ratios of Ni^{2+}/Fe^{3+} have been successfully produced using an appropriate solvothermal router. Physicochemical properties of all samples were characterized using X-ray diffraction (XRD), Raman, field emission scanning electron microscopes (FE-SEM), fourier-transform infrared spectroscopy (FT-IR), N_2 adsorption-desorption analysis, X-ray photoelectron spectroscopy (XPS), ultraviolet-visible diffuse reflectance spectra (UV-Vis DRS), and photoluminescence spectra (PL). The photocatalytic degradation performances of the photocatalysts were evaluated in the decomposition of rhodamine B (RhB) under a compact fluorescent daylight lamp. From XRD, IR, XPS, and Raman results, with the presence of mixed ion Fe^{3+} and Ni^{2+}, MIL-88B (MIL standing for Materials of Institut Lavoisier) crystals based on the mixed metal Fe_2NiO cluster were formed, while MIL-53(Fe) was formed with the presence of single ion Fe^{3+}. From UV-Vis DRS results, Ni/Fe-MOF samples exhibited the absorption spectrum up to the visible region, and then they showed the high photocatalytic activity under visible light irradiation. A Ni/Fe-MOF sample with a Ni^{2+}/Fe^{3+} molar ratio of 0.3 showed the highest photocatalytic degradation capacity of RhB, superior to that of the MIL-53(Fe) sample. The obtained result could be explained as a consequence of the large surface area with large pore volumes and pore size by the Ni^{2+} incorporating into the MOF's structure. In addition, a mixed metal Fe/Ni-based framework consisted of mixed-metal cluster Fe_2NiO with an electron transfer effect and may enhance the photocatalytic performance.

Keywords: photocatalytic decomposition of rhodamine B; MIL-53(Fe); Ni/Fe-MOF; visible light irradiation

1. Introduction

Metal-organic frameworks (MOFs), a new class of high surface area and crystalline porous materials, assemble with metal clusters and organic bridging ligands [1]. These materials have received considerable attention in recent years due to their high resistance, high surface area, large pore volume, low density, and easily tunable framework. Among the MOFs, MIL-53(Fe) 88B (MIL standing for Materials of Institut Lavoisier) have attracted extensive interest for applications in gas storage [2,3], adsorption and separation of heavy metal [4], sensors [5], and in the biomedical field such as for drug delivery [6].

Recently, to eliminate organic dyes, many approaches have been suggested including adsorption [7–10] and photodegradation [11–13]. However, the latter is of interest because this process could decompose organic dyes to CO_2, H_2O, and harmless inorganics, while the adsorption process is only capable of removing dyes from water media. MIL-53(Fe) as a catalyst carrier or modification of MIL-53(Fe) as a catalyst for chemical reactions has received research attention [14]. MIL-53(Fe) has the chemical formula of $Fe^{III}(OH)(O_2C-C_6H_4-CO_2) \cdot H_2O$, which consists of FeO_6 octahedral chains connected to benzene dicarboxylate (BDC) anions, forming a three-dimensional network with a large volume and high surface area [2,14,15]. The FeO_6 octahedral chains have the potential to act as a Lewis acid in many organic reactions [16]. Recently, MIL-53(Fe) with the potential use of FeO_6 octahedral chains has received much attention in photocatalytic degradation of many organic dyes, such as methylene blue [11,13,17], rhodamine B (RhB) [14,16,17], and p-nitrophenol [14], and has given good decomposition results. Therefore, this is a possible application direction of MIL-53(Fe) in the removal of organic dyes.

Fe-based MOFs materials have been reported as an effective photocatalyst for decomposition of organic dyes under visible light irradiation [18–22]. However, their photocatalytic performance is not as expected because of the fast recombination of photogenerated holes (h^+) and electrons (e^-), resulting in the lack of h^+ for degradation dyes [13]. To address this, various approaches have been proposed to depress the recombination process. For example, inorganic oxidants (e.g., H_2O_2, $KBrO_3$, and $(NH_4)_2S_2O_8$), which act as electron acceptors, was introduced in the photocatalytic processes, significantly enhancing the photocatalytic effect of these materials. According to research by Yuan et al. [13], H_2O_2 is an efficient electron acceptor in the photocatalytic decomposition process of organic pigments by MIL-53(Fe) under visible light irradiation. Another approach that has been developed to enhance the photocatalytic performance of MiL-53(Fe) is the designed synthesis of composite photocatalysts containing MOFs materials such as CdS/MIL-53(Fe) [23], Ni-MOFs@GO [24], Fe_3O_4/MIL-53(Fe) [14], and Fe_2O_3/MIL-53(Fe) [25]. In addition, MIL-53(Fe) that has been doped or combined with one or more metals have also attracted much attention in recent years [26–29]. For this study, Qiao Sun et al. modified the MIL-53(Fe) by adding Mn, Co, and Ni metal into the framework of MIL-53(Fe) material, which exhibited excellent catalytic performance in liquid-phase degradation of phenol [30]. Various rare-earth or transition metals that modify MOFs structures have recently been reported such as three-dimensional Ln(III)–Zn(II) heterometallic coordination polymers [31], Fe substituted Cr MIL-101 [32], Ag-doped MOF-like organotitanium polymer (Ag@NH2-MOP(Ti)) [33], Ti-doped UiO-66 [34], Eu substituted Fe MIL-53 [35], and Zn-Ln coordination polymers (Ln = Nd, Pr, Sm, Eu, Tb, Dy) [36].

In this work, we report the synthesis of Ni/Fe-MOF with different Ni^{2+}/Fe^{3+} molar ratios using the solvothermal route and their application for the degradation of RhB solution under visible light irradiation using a 40 W compact fluorescent lamp. To illustrate our method for the synthesis of Ni/Fe-MOF, we have selected the preparation of the MIL-53(Fe) structure, which consists of FeO_6 octahedral chains connected to BDC anions. Thanks to the presence of Ni^{2+} ions in the reaction solution, MIL-88B crystals were formed with neutral mixed-metal clusters (Fe_2NiO) connected via BDC anions. This structure is similar to the MIL-88B structure consisting of the trinuclear oxo-centered iron cluster (Fe_3O) [27,28]. However, our bimetallic metal MOF products were expected to exhibit an excellent adsorption capacity and photocatalytic activity in comparison to the original single metal MOFs.

The advantage of selecting MOF material containing Fe and Ni is due to the low cost, non-toxicity, and natural abundance of these two transition metal oxides. In addition, the MOF material is also capable of improving the separation efficiency of electron–hole pairs when Ni is incorporated into the structure of materials [37,38]. The structure, morphology, and optical properties of the obtained photocatalysts have been characterized using X-ray diffraction (XRD), Raman, field emission scanning electron microscopes and energy-dispersive X-ray spectrometer (FE-SEM/EDS), fourier-transform infrared spectroscopy (FT-IR), N_2 adsorption-desorption analysis, X-ray photoelectron spectroscopy (XPS), ultraviolet-visible diffuse reflectance spectra (UV-Vis DRS), photoluminescence (PL) spectra and nitrogen physisorption measurements (BET). Besides, to obtain the optimal reaction conditions for the RhB photodecomposition, the effect of the initial RhB concentration and pH on the degradation of RhB was also investigated in detail.

2. Results and Discussion

2.1. Physical Properties of MIL-53(Fe) and Ni-Doped MIL-53(Fe)

2.1.1. XRD Analysis

Figure 1 presents the XRD diffraction patterns of the MIL-53(Fe) and Ni/Fe-MOF samples isolated from dimethylformamide (DMF) and H_2O. In patterns of MIL-53(Fe) samples (Figure 1A, curve a), the main diffraction peaks that appeared at 2θ of 9.1°, 9.4°, 14.1°, 16.5°, and 18.8° are similar to those previously reported for MIL-53(Fe) isolated from DMF [2,11,39]. In patterns of Ni/Fe-MOF samples (Figure 1A, curves b–e), the main diffraction peaks that appeared around 2θ of 7.3°, 8.9°, 9.3°, 9.9°, 16.8°, 18.7°, 17.7°, 20.1°, and 21.9° are similar to those previously reported for MIL-88B isolated from DMF. Notably, the diffraction peak at a 2θ of 7.3° observed in the XRD patterns of Ni/Fe-MOF samples increased in intensity as the molar ratio of Ni^{2+}/Fe^{3+} increased from 0.1 to 0.7. With the presence of Ni^{2+} in the reaction solution, MIL-88B crystals were made up and the crystallinity of the material increased. This observation might be attributed to the fact that the structure formation of Ni/Fe-MOF was significantly influenced by the presence of Ni^{2+} in the reaction solution. In addition, no other diffraction peak associated with nickel oxides, iron oxides, or other impurities could be detected, demonstrating the high purity of the samples.

XRD patterns of the MIL-53(Fe) and Ni/Fe-MOF samples isolated from H_2O (Figure 1B) showed the rugged background and weak intensities; however, the main diffraction peaks still maintained the same structure as in Reference [4]. The difference in XRD patterns of samples isolated from DMF and water may attribute to the breathing behavior of MIL-53(Fe) and MIL-88B, which has been well documented by Alhanami et al. [15]. Moreover, MIL-53(Fe)·H_2O sample essentially shows a noncrystalline phase similar to those for MIL-53(Fe)·DMF. They can be explained by the effect of the synthesis temperature on the structure formation of MIL-53(Fe). Pu et al. demonstrated that iron ion and H_2BDC could not coordinate successfully under a low temperature (100 °C), and therefore the MIL-53(Fe) crystal structure could not fully develop [40]. However, the Ni/Fe-MOF samples still show a high crystalline phase under low synthesis temperatures. Again, these results indicate that the presence of a mixed metal ion (Ni^{2+} and Fe^{3+} ion) did have a significant influence on the formation of Ni/Fe-MOF crystal structure, in which a Ni^{2+} and Fe^{3+} ion can coordinate with H_2BDC to form MIL-88B crystals instead of MIL-53(Fe) crystals.

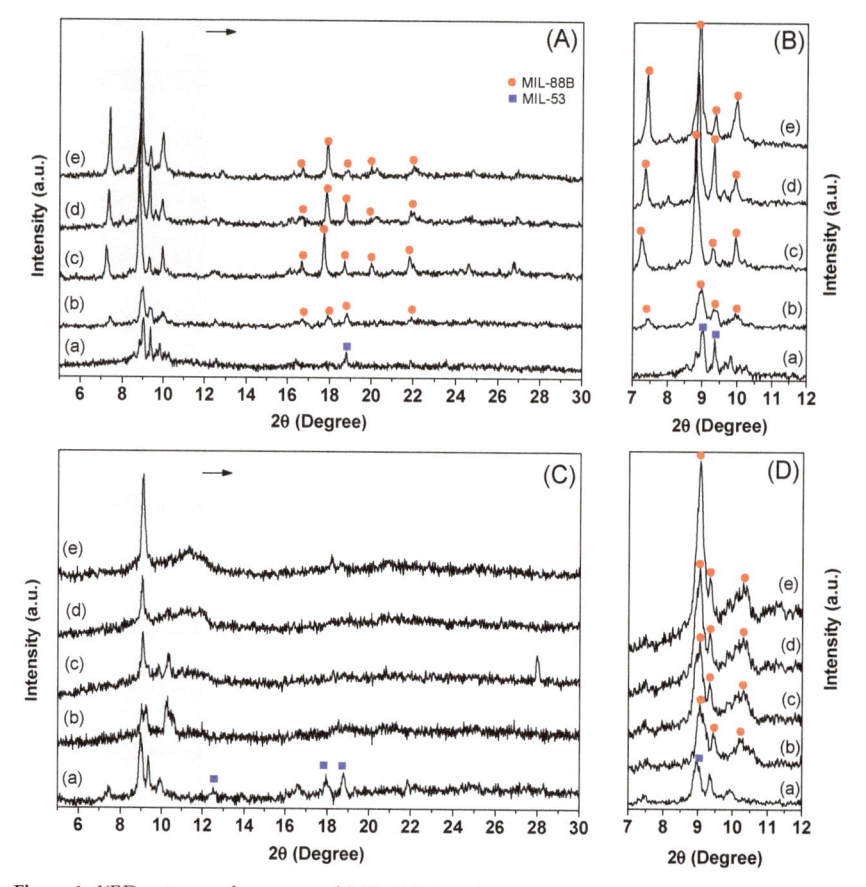

Figure 1. XRD patterns of as-prepared MIL-53(Fe) and Ni-MIL-53(Fe) crystals isolated from DMF (**A,B**) and H$_2$O (**C,D**): MIL-53(Fe) (a), Ni/Fe-MOF-0.1 (b), Ni/Fe-MOF-0.3 (c), Ni/Fe-MOF-0.5 (d), and Ni/Fe-MOF-0.7 (e).

2.1.2. FT-IR Spectra

FTIR spectroscopic studies were performed for all samples in the wave range of 400–4000 cm^{-1}, as shown in Figure 2. As shown in Figure 2A,C, strong vibrational bands around 1657, 1601, 1391, 1017, and 749 cm^{-1}, which are attributed to υ(C=O), υ_{as}(OCO), υ_{s}(OCO), υ(C–O), and δ(C–H) vibrations confirms the presence the bridge coordination mode of metal carboxylates in the MOF structures [4,25,30]. No band at 1700 cm^{-1} was found, implying no free H$_2$BDC [27]. The band characteristics of DMF (1657 cm^{-1}) and H$_2$O (3387 cm^{-1}) were present in the samples MIL-53(Fe)·DMF, Ni/Fe-MOF·DMF, MIL-53(Fe)·H$_2$O and Ni/Fe-MOF-x·H$_2$O, respectively [27].

At lower frequencies (Figure 2B), vibrational bands around 750 cm^{-1}, 690 cm^{-1}, and 660 cm^{-1} represent the C–H vibration, C=C stretch, OH bend, and OCO bend, respectively, were found, implying the presence of the vibrations of the organic ligand BDC [27]. Figure 2B also shows that the strong band at 547 cm^{-1} in all samples could be attributed to Fe–O vibrations or Ni–O vibrations [41]. The band around 625 cm^{-1} belongs to the Fe$_3$O vibration, which was observed in MIL-53(Fe) and Ni-Ni/Fe-MOF-0.1 samples. The weak band around 720 cm^{-1} is related to the Fe$_2$NiO vibration, which was observed in Ni/Fe-MOF-x samples [27]. These results reaffirmed that Ni^{2+} and Fe^{3+} ions can coordinate with H$_2$BDC to form MIL-88B crystals.

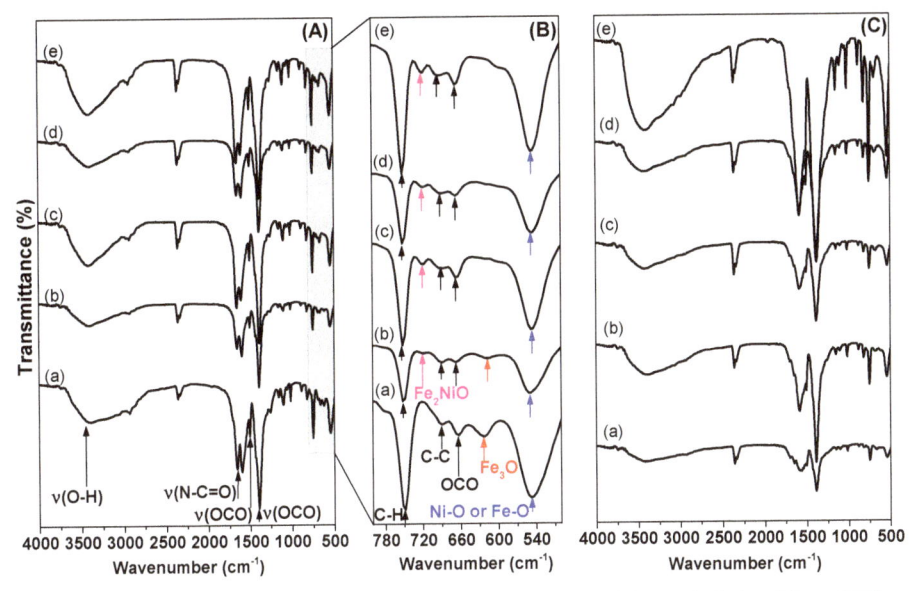

Figure 2. FT-IR spectra of as-prepared MIL-53(Fe) and Ni/Fe-MOF crystals isolated from DMF (**A**,**B**) and H$_2$O (**C**): MIL-53(Fe) (a), Ni/Fe-MOF-0.1 (b), Ni/Fe-MOF-0.3 (c), Ni/Fe-MOF-0.5 (d), and Ni/Fe-MOF-0.7 (e).

2.1.3. Raman Spectra

Samples were analyzed using Raman spectroscopy using an excitation wavelength at 633 nm and spectra recorded at a wavenumber range of 100–900 cm^{-1}, as shown in Figure 3. According to previous studies, the BDC bridge in MOFs has Raman-active modes: the symmetric vibration modes (vs. (COO)) and asymmetric vibration (vas (COO)) of the carboxylate group (1445 cm^{-1} and 1501 cm^{-1}), the vibration of the C–C bond between the benzene ring and the carboxylate group (1140 cm^{-1}), and the external plane deformation of the C–H link (865 cm^{-1} and 630 cm^{-1}) [28]. As seen in Figure 3, the presence of a BDC linker was also observed in all samples, and no Raman signals corresponding to nickel oxides, iron oxides, or other impurities were found on any of the samples, which is consistent with the results of the XRD patterns. Notably, the Raman signal corresponding to the symmetric vibration (vs. (OCO)) of the carboxylate group showed a shift to a lower wavenumber and the peak split into two peaks corresponding to an increase of the Ni^{2+}/Fe^{3+} molar ratio. This result was due to the change in the charge distribution in the organic bridge when they were coordinated with different metal ions (Figure 3B). Ionic Ni^{2+} has a smaller nuclear charge and a larger ionic radius than Fe^{3+} ($r_{Ni^{2+}}$ = 0.69 Å and $r_{Fe^{3+}}$ = 0.55 Å) [42]. Therefore, Ni^{2+} creates a weaker coordinated link with the OCO group on the organic bridge than Fe^{3+}, thus the symmetric vibration (vs. (OCO)) of the carboxylate group when forming coordinated bonds with Ni^{2+} moves to a lower wavenumber than Fe^{3+} [43]. This result is commensurate with the XRD and IR results for Ni/Fe MOF.

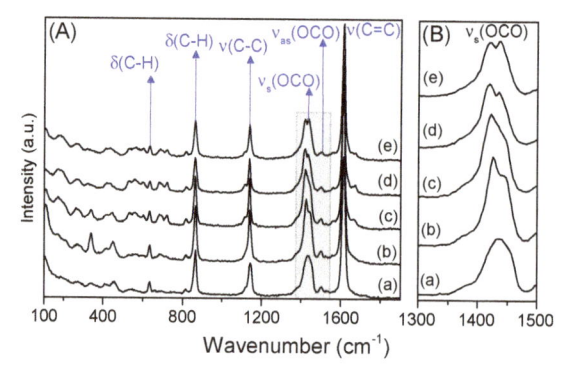

Figure 3. Raman spectra of as-prepared MIL-53(Fe) and Ni/Fe-MOF crystals isolated from DMF (**A**): MIL-53(Fe) (a), Ni/Fe-MOF-0.1 (b), Ni/Fe-MOF-0.3 (c), Ni/Fe-MOF-0.5 (d), and Ni/Fe-MOF-0.7 (e), and enlarged Raman spectra around 1450 cm^{-1} (**B**).

2.1.4. FE-SEM/EDS Analysis

Figure 4 displays SEM images and EDS spectra of the as-prepared MOF samples. As shown in Figure 4, the morphologies and shapes of MOF samples varied according to the molar ratio of Ni^{2+}/Fe^{3+}. MIL-53(Fe) sample mostly had amorphous nanoparticles (Figure 4(a1,a2)), which is in good agreement with the results of XRD patterns with a poor crystallinity. When the molar ratio of Ni^{2+}/Fe^{3+} was set to 0.1, the crystals of Ni/Fe-MOF-0.1 were not homogeneous with different shapes and sizes (Figure 4(b1,b2)). A mixture of octahedral and hexagonal bipyramidal shapes, and nanoparticles, were perceived when the molar ratio of Ni^{2+}/Fe^{3+} (0.3–0.7) was increased further. However, these octahedral and hexagonal bipyramidal shapes collapsed with cracks on the crystal surface. These results, along with the XRD, IR, and Raman results above, indicate that a mixed-metal Ni/Fe-MOF was successfully synthesized using the solvothermal method.

Moreover, to confirm the molar ratio of Ni^{2+}/Fe^{3+} in the Ni/Fe-MOF samples in comparison to the theoretical value, EDS was also conducted. The result from the EDS spectrum of the obtained MIL-53(Fe) sample (Figure 4(a3)) showed the coexistence of C, O, Fe, and Cl. The presence of Cl may have been due to the $FeCl_3$ precursor, further confirming that the MIL-53(Fe) crystal structure could not fully develop at a low temperature (100 °C). The EDS spectra of the Ni/Fe-MOF samples (Figure 4(b3,c3,d3,e3)) revealed that these samples contained C, O, Fe, and Ni. However, the existence of Cl was still observed in the Ni/Fe-MOF-0.1 sample. The molar ratio of Ni^{2+}/Fe^{3+} of Ni/Fe-MOF-0.1, Ni/Fe-MOF-0.3, Ni/Fe-MOF-0.5, and Ni/Fe-MOF-0.7, obtained using EDS analysis, was 0.16, 0.30, 0.48, and 0.66, respectively. In addition, the map of Fe, O, C, and Ni is shown in Figure S1, which indicates that they were uniformly distributed over the MOF surface.

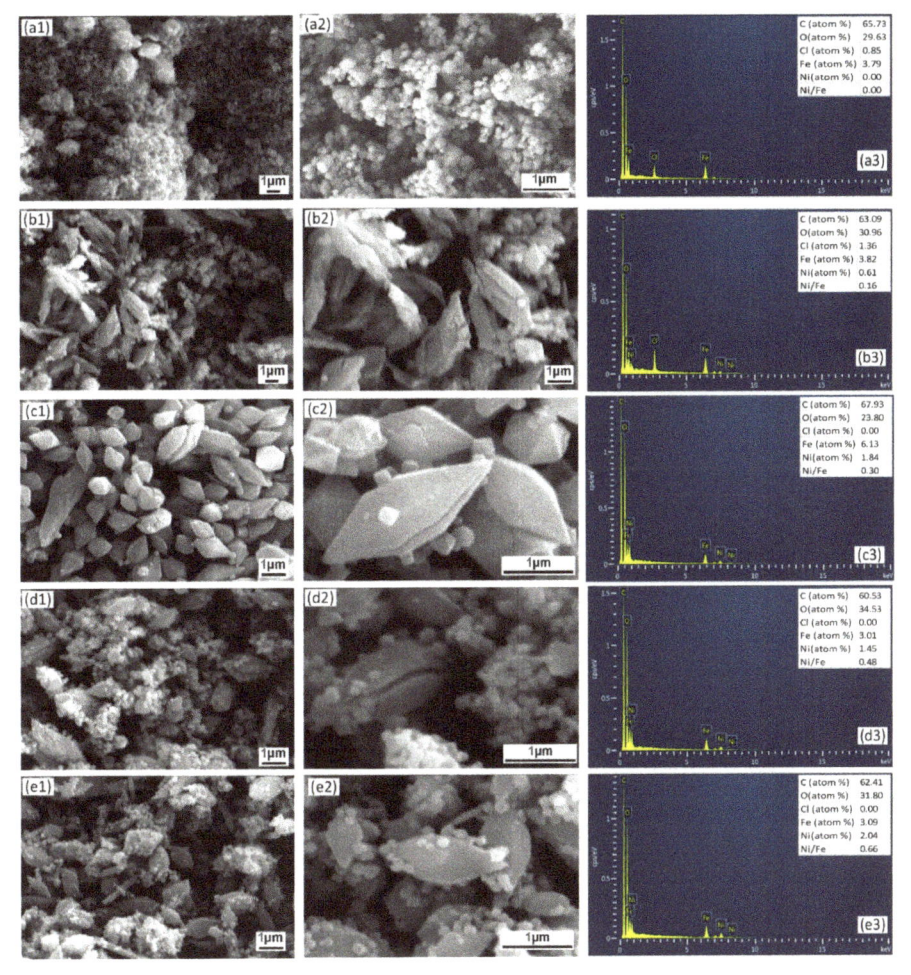

Figure 4. SEM images (1, 2) and EDS patterns (3) of as-prepared MIL-53(Fe) and Ni/Fe-MOF crystals isolated from DMF: MIL-53(Fe) (**a**), Ni/Fe-MOF-0.1 (**b**), Ni/Fe-MOF-0.3 (**c**), Ni/Fe-MOF-0.5 (**d**), and Ni/Fe-MOF-0.7 (**e**).

2.1.5. XPS Spectra

To analyze the chemical states of Ni and Fe in the Ni/Fe MOF structure, XPS spectroscopy was carried out. As illustrated in Figure 5A, the wide-scan XPS spectra of MIL-53(Fe)·H_2O possesses the characteristic peaks of C, O, Fe, and Cl, while Ni/Fe-MOF-0.3·H_2O contained C, O, Fe, and Ni. Based on the XPS analysis, the Ni/Fe-MOF-0.3 had a surface molar ratio of Ni^{2+}/Fe^{3+} of 0.26, which approximates the EDS results above. Besides, N was not detected in either sample, indicating that the DMF solvent was sufficiently eliminated from the MOFs.

Figure 5B shows the C 1s XPS spectra of MIL-53(Fe)·H_2O and Ni/Fe-MOF-0.3·H_2O samples. Both spectra were fitted into three peaks at a binding energy (BE) of 285.01, 288.9, and 291.7 eV, which could be assigned to the carbon components on the phenyl and the carboxylate groups of the BDC linkers [30,40,44–46]. The O 1s XPS spectra (Figure 5C) could also be fitted into three peaks, which are (i) the peak at 533.8 eV corresponding to the O components on C=O/H_2O, (ii) the peak at 532.3 eV attributed to the O components on the BDC linkers, and (iii) the peak at 530. 5 eV was assigned to the

O components on the Fe–O bonds (for MIL-53(Fe) sample) or Fe$_2$NiO clusters (for Ni/Fe-MOF-0.3 sample). These results further confirmed the coordination between the metal ion (Ni^{2+} and/or Fe^{3+}) and BDC linkers, which is commensurate with the XRD, IR, and Raman results above.

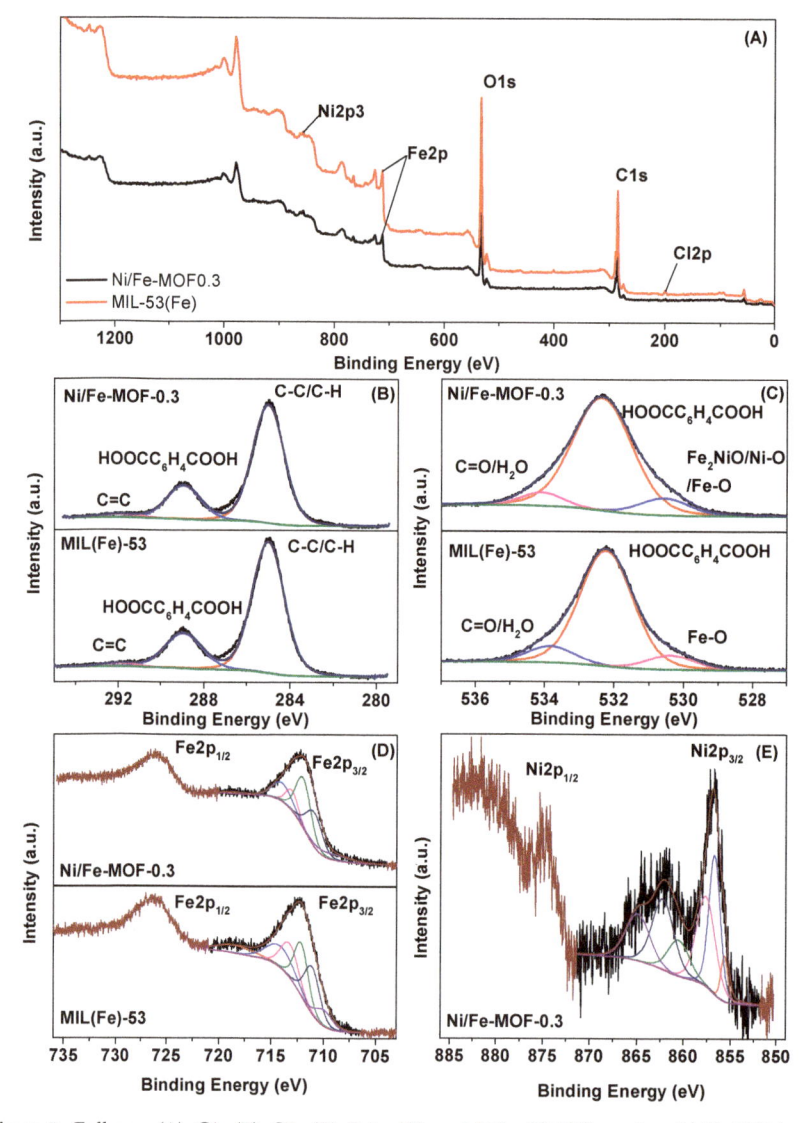

Figure 5. Full scan (**A**), C1s (**B**), O1s (**C**), Fe2p (**D**), and Ni2p (**E**) XPS spectra of MIL-53(Fe) and Ni/Fe-MOF-0.3.

The Fe 2p high-resolution XPS spectrum of MIL-53(Fe) sample (Figure 5D) displays two main peaks that were indexed to Fe 2p1/2 (712.4 eV) and Fe2p3/2 (726.1 eV). The splitting energy of the 2p doublet was 13.7 eV, implying that the valence state of Fe was +3 [4,23,44]. Similarly, the valence state of Fe in the Ni/Fe MOF structure was also +3 because the splitting energy between Fe 2p1/2 (712.9 eV) and Fe 2p3/2 (726.2 eV) was 13.3 eV. To further confirm the valence state of Fe in both of these samples, the Fe 2p3/2 peak was fitted into six peaks including Gupta and Sen (GS) multiples, surface structures, and shake-up-related satellites [28,47,48]. The fitting results, as shown in Figures S6 and S7, were indexed well with Fe^{3+} GS multiplets, which indicated that the valence state of Fe in the MIL-53(Fe) and Ni/Fe MOF structure was +3. In the high-resolution XPS spectrum of Ni 2p (Figure 4e), we observed the BE of the Ni 2p3/2 (857.2 eV) and Ni 2p1/2 (874.8 eV) core-level peaks with the doublet separation of 17.6 eV, implying that the valence state of Ni was +2 [49,50].

2.1.6. N_2 Adsorption/Desorption

The specific surface area and porous structure of MIL-53(Fe) and Ni/Fe-MOF crystals isolated from DMF and H_2O were determined using N_2 adsorption–desorption isotherms at 77 K. The N_2 adsorption–desorption isotherms, as shown in Figure 6A, displayed an intermediate mode between type I and type IV, which was associated with mesoporous and microporous materials, respectively [51]. The Brunauer–Emmett–Teller (BET) surface area, pore volume, and pore width of MIL-53(Fe) and Ni/Fe-MOF-0.3 samples are shown in Table 1. The MIL-53(Fe)·H_2O, MIL-53(Fe)·DMF, Ni/Fe-MOF-0.3.H_2O, and Ni/Fe-MOF-0.3.DMF had specific surface areas of 158, 300, 247, and 480 m^2/g, respectively (Table 1). The mesopore size distribution curve of samples calculated using the Barrett–Joyner–Halenda (BJH) model is shown in Figure 6B. The MIL-53(Fe)·H_2O and MIL-53(Fe)·DMF sample was non-porous, whereas Ni/Fe-MOF-0.3·H_2O, and Ni/Fe-MOF-0.3·DMF showed a pore size centered at about 3.8 nm and 21.4 nm, respectively. Therefore, compared with MIL-53(Fe), Ni/Fe-MOF-0.3 showed a higher value in the specific surface areas. In addition, the higher surface area and micropore volume for samples isolated from DMF, as compared with samples isolated from H_2O, was due to the reversible breathing behavior of these materials, which was dependent on the molecule present inside their pores, where the pores were opened in the presence of DMF and closed in the presence of H_2O [27,28,52]. The formation of porous material for Ni/Fe-MOF-0.3 could be explained by the formation of Fe_2NiO cluster in the Ni/Fe-MOF structure, which could affect the reversible breathing behavior of these materials. MIL-88B(Fe) crystals with trinuclear metal clusters were known as non-porous materials due to the need for compensating the anion inside their porous system [28,53]. Do and coworkers demonstrated that MOF structure with the presence of Fe_2NiO cluster as nodes in the MIL-88B framework avoids the compensating anion [27,28], which results in the formation of porous material for Ni/Fe-MOF-0.3. In addition, the cracks on the crystal surface of Ni/Fe-MOF-0.3 (Figure 4) could also partly create the characteristics of microporous or mesoporous materials for this sample.

Table 1. Specific surface area and porosity of MiL-53(Fe) and Ni/Fe-MOF samples.

Samples	Specific Surface Area (m^2/g)	Micropore Volume ($\times 10^{-3}$ cm^3/g)	Mesopore Volume ($\times 10^{-3}$ cm^3/g)	Average Pore Width (nm)
MIL-53(Fe)·DMF	300	128	97	13
MIL-53(Fe)·H_2O	158	65	59	11
Ni/Fe-MOF-0.3·DMF	480	212	128	8
Ni/Fe-MOF-0.3·H_2O	247	94	271	13

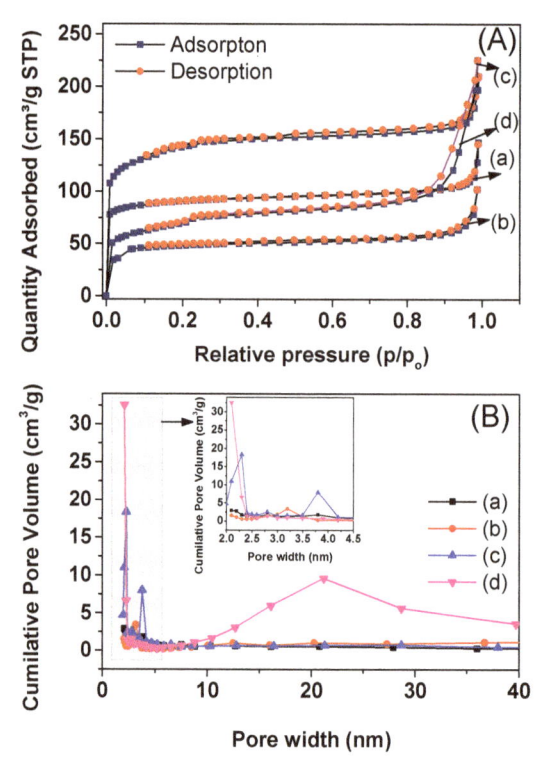

Figure 6. N_2 adsorption–desorption isotherms (**A**) and pore size distributions (**B**) of the synthesized samples: MIL-53(Fe)·DMF (a), MIL-53(Fe)·H_2O (b), Ni/Fe-MOF-0.3·DMF (c), and Ni/Fe-MOF-0.3·H_2O (d).

2.1.7. UV-Vis Spectra

The light absorption properties of the material were studied through the UV-Vis-DRS spectra. The UV-Vis-DRS spectrum of the material is shown in Figure 7. For washing samples with DMF (Figure 7A), MIL-53(Fe)·DMF gave strong absorption bands in the wavelength range of 200 to 400 nm. The strong absorption bands at 256 to 310 nm could be due to the transfer of the charge from the oxygen center of the organic bridge to the metal center in the octahedral FeO_6 structure [17,54]. The band at 350 to 500 nm was due to the shift of d–d (6A1g → 4A1g + 4Eg (G)) of Fe^{3+} in the MIL-53(Fe) structure [14,27]. The main absorption edge (λ, nm) of the MIL-53(Fe)·DMF was 478 nm, corresponding to the bandgap energy E_g = 2.59 eV (E_g = 1240/λ). This result is in accordance with previous reports [44,55]. When the MIL-53(Fe) was modified with Ni, the material have the decreased absorption in the wavelength range from 200 to 500 nm, and the absorption spectrum extended in the range from 250 to 800 nm, so it was difficult to determine the absorption of the material accurately. When the material was washed with water (Figure 7B), the modified material had an increased absorption in the wavelength range from 200 to 400 nm, and the absorption intensity was higher and broader in the visible light region as compared to the modified sample washed with DMF. As the material was washed with water, there was a structural change between the large pore and the narrow pore caused by the "breathing" effect when the material absorbed the water molecules inside the pore. This phase transformation of the structure led to a change in the electronic structure [56], and subsequently, a change in the absorption spectrum of the material and decreasing E_g. For Ni/Fe-MOF-0.1·H_2O, Ni/Fe-MOF-0.3·H_2O, and Ni/Fe-MOF-0.5·H_2O samples, the absorption intensity in the visible light region and the absorption band of the material shifted to a wavelength longer than for MIL-53(Fe)·H_2O. As absorption in the visible light increased, the visible light energy could be used more efficiently, thus contributing to the increased photocatalytic efficiency of the material. The absorption edges of MIL-53(Fe)·H_2O, Ni/Fe-MOF-0.1·H_2O, Ni/Fe-MOF-0.3·H_2O, Ni/Fe-MOF-0.5·H_2O, and Ni/Fe-MOF-0.7·H_2O were 504, 553, 532, 513, and 516 nm (Figure S2), corresponding to the optical bandgap of 2.46, 2.24, 2.33, 2.42, and 2.40 eV, respectively. These results provided a potential photoreactivity of MIL-53(Fe) and Ni/Fe-MOF samples in the visible light range.

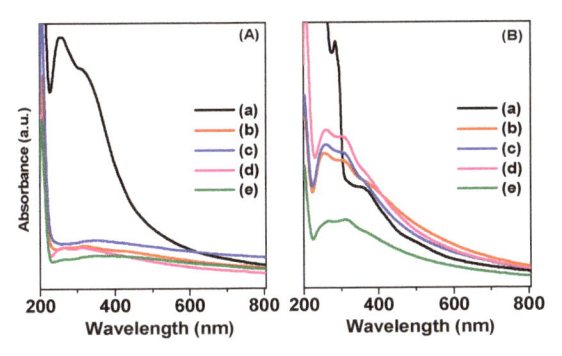

Figure 7. UV-Vis DRS spectra of as-prepared MIL-53(Fe) and Ni-MIL-53(Fe) crystals isolated from DMF (**A**) and H_2O (**B**): MIL-53(Fe) (a), Ni/Fe-MOF-0.1 (b), Ni/Fe-MOF-0.3 (c), Ni/Fe-MOF-0.5 (d), and Ni/Fe-MOF-0.7 (e).

2.1.8. PL Spectroscopy

PL spectra of MIL-53(Fe) and Ni/Fe-MOF samples were recorded at room temperature and are shown in Figure 8. When the MIL-53(Fe) sample was excited by a 320 nm laser, its emission spectrum showed a strong emission range of 350 to 500 nm and a weak emission range of 570 to 750 nm. In comparison, the intensity of Ni/Fe-MOF samples was significantly lower than that of the MIL-53(Fe) sample because of the presence of the Ni_2FeO cluster in the structure of the Ni/Fe-MOF crystal. These results demonstrated that electron–hole recombination could be inhibited in the Ni/Fe-MOF, resulting in the improvement of photocatalytic performance. PL spectra, along with the UV-Vis DRS result, could satisfy the prerequisite for visible-light photocatalysis.

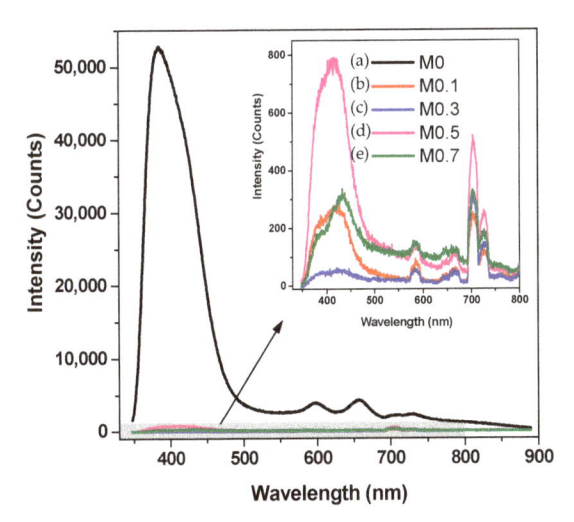

Figure 8. PL spectra of as-prepared MIL-53(Fe) (a), Ni/Fe-MOF-0.1 (b), Ni/Fe-MOF-0.3 (c), Ni/Fe-MOF-0.5 (d), and Ni/Fe-MOF-0.7 (e).

2.2. Photocatalytic Activities

2.2.1. RhB Removal by MIL-53(Fe) and Ni-MIL-53(Fe)

The photocatalytic activities of MIL-53(Fe) and Ni/Fe-MOF-x photocatalysts were evaluated in the liquid-phase photodegradation of RhB dye under visible light irradiation. Figure 9 displays the changes of RhB concentrations via adsorption and photocatalytic degradation under different experimental conditions. As shown in Figure 9, a negligible degradation of RhB concentrations was observed in the several blank runs including RhB/H_2O_2/Dark, RhB/H_2O_2/Light, and RhB/Dark systems, proving the stability property of RhB under visible light irradiation of compact fluorescent light. Also, as shown in Figure 9A, after 180 min adsorption (in the dark), 16% and 51% RhB were removed in the presence of MIL-53(Fe) (MIL-53(Fe)/Dark system) and Ni/Fe-MOF-0.3 (Ni/Fe-MOF-0.3/Dark system), respectively. The higher adsorption capacity of the Ni/Fe-MOF-0.3 sample was due to its higher surface area (247 m^2/g for Ni/Fe-MOF-0.3 and 158 m^2/g for MIL-53(Fe)). In addition, there was no significant difference in the removal of RhB concentration in the two adsorption experiments with the presence of H_2O_2 (MIL-53(Fe)/H_2O_2/Dark and Ni/Fe-MOF-0.3/H_2O_2/Dark systems) and the absence of H_2O_2 (MIL-53(Fe)/Dark and Ni/Fe-MOF-0.3/Dark systems). Therefore, our photocatalytic experiments do display the presence of a Fenton reaction.

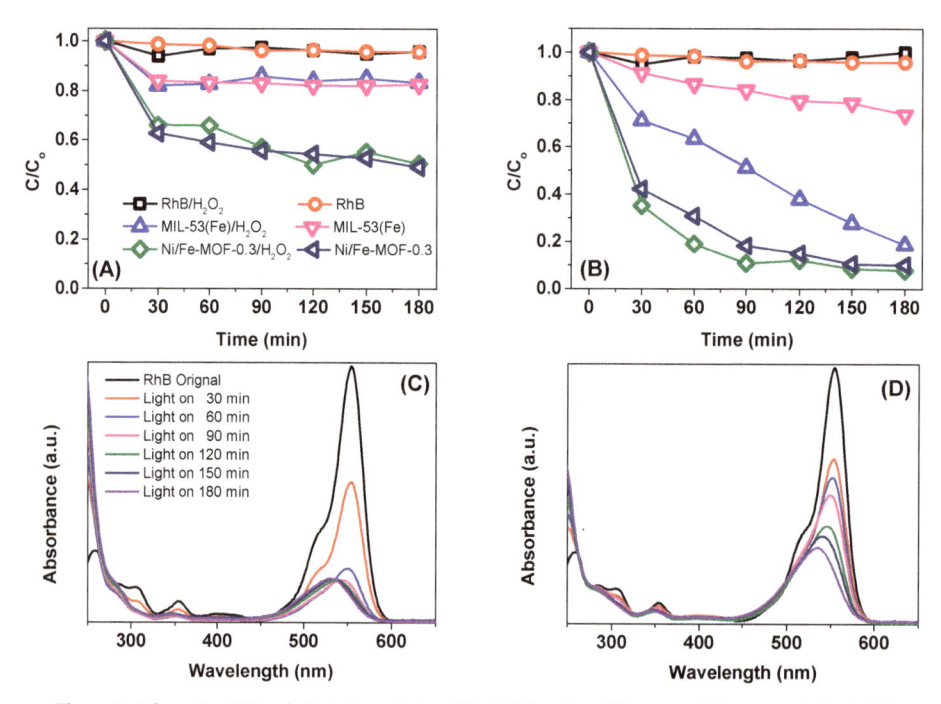

Figure 9. Adsorption (**A**) and photodegradation (**B**) of RhB under different conditions over MIL-53(Fe) and Ni/Fe-MOF-0.3, and UV-Vis spectral of RhB solution separated from the Ni/Fe-MOF-0.3/Light/H_2O_2 catalytic system (**C**) and MIL-53(Fe)/Light/H_2O_2 catalytic system (**D**).

Under visible light irradiation, the presence of MIL-53(Fe) could enhance the degradation efficiency of RhB up to 81.46% using a photolysis process in MIL-53(Fe)/Light/H_2O_2 catalytic system (Figure 9B). For the Ni/Fe-MOF-0.3/Light/H_2O_2 catalytic system, the degradation efficiency of RhB was remarkably enhanced where about 91.14% RhB removal was achieved (Figure 9B). The higher photocatalytic activity of the Ni/Fe-MOF-0.3 sample as compared with MIL-53(Fe) could also be indicated by the change of the UV-Vis absorption spectra of the solution in the course of the RhB degradation (Figure 9C,D). As seen in Figure 9C,D, the primary absorption band, which could be attributed to RhB, shifted from 554 to 500 nm in a step-wise manner. This change could be reasonably assigned to the removal of ethyl groups one by one in this reaction, which is in good agreement with the previous literature. The photodegradation of RhB over MIL-53(Fe) and Ni/Fe-MOF-0.3 photocatalysts approximately followed a pseudo-first-order kinetics model: $\ln(C_o/C) = k_{obs}t$ [57–59]. The presence of Ni/Fe-MOF-0.3 promoted the photodegradation rate; the rate constants were 8.88×10^{-3} min^{-1} for MIL-53(Fe) and 11.15×10^{-3} min^{-1} for Ni/Fe-MOF-0.3.

To investigate the role of H_2O_2 on the photocatalytic performance of MIL-53(Fe) and Ni/Fe-MOF photocatalysts, the photocatalytic processes with the presence and absence of H_2O_2 were conducted in parallel (Figure 9B). After 180 min of irradiation, the degradation rate of RhB over MIL-53(Fe)/Light/H_2O_2 and MIL-53(Fe)/Light process was 81.46% and 27.60%, respectively. Only MIL-53(Fe) with the absence of H_2O_2 exhibits the low efficiency of RhB photodegradation due to the fast electron-hole recombination, which is in good agreement with the previous literature [13,17]. For the MIL-53(Fe)/H_2O_2/Light process, H_2O_2 acted as an electron acceptor, resulting in the suppression of charge recombination; therefore, the rate for RhB decomposition could be significantly enhanced, as was demonstrated by Du et al. [13]. Similarly, Ai et al. also showed that the enhancement of MI-53(Fe) photocatalytic performance could be due to the synergistic effects of the combination of MIL-53(Fe) and H_2O_2 under visible light irradiation [17]. Interestingly, the effect of H_2O_2 on the photocatalytic performance of the Ni/Fe-MOF photocatalyst showed a considerable difference. The Ni/Fe-MOF sample could degrade more than 90% of the initial RhB content regardless of the presence or absence of H_2O_2.

The superior catalytic performance of the Ni/Fe-MOF sample could be explained by the formation of the mixed metal cluster Fe_2NiO in the Ni/Fe-MOF framework. According to recent reports, the Fe-based framework (MIL-101, MIL-100, MIL-88, and MOF-235), containing single metal cluster Fe_3-μ_3-oxo clusters with small particle sizes, are proposed as a visible light photocatalyst [44,60–63]. The reaction mechanism of these materials have been reported based on semiconductor theory and previous reports [61–64]. Particularly, when the surface of MOFs material absorbs photons ($E_{photons} \geq E_g$), the electrons (e^-) in the valence band (VB) will be excited to the conduction band (CB), leaving the holes (h^+) in the VB. These photogenerated e^-–h^+ pairs may be further involved in the following three processes: (i) successfully migration to the surface of MOFs, (ii) being captured by the defect sites in bulk and/or on the surface region of semiconductor, and (iii) recombining and releasing the energy in the form of heat or a photon. Then, the h^+ can accept electrons and induce water molecules to generate hydroxyl radicals (•OH), which exhibit a high oxidation ability to decompose the organic dyes. However, there is a recombination of excessive electrons and holes, resulting in the restricted photocatalytic activity of this material. In our study, mixed a metal Fe/Ni-based framework that consists of a mixed-metal cluster Fe_2NiO with electron transfer effect may enhance the photocatalytic performance [45,61,65].

Besides, a mixed metal Fe/Ni-based framework that consists of the mixed-metal cluster Fe_2NiO possesses large pores and a high surface area, as compared with a single metal Fe-based framework; therefore, Ni/Fe-MOF exhibited a high adsorption capacity of RhB and high photocatalytic activity in RhB degradation. XRD patterns of Ni/Fe-MOF-0.3 before and after reactions were shown in Figure S3 (SI file). As shown in Figure S3, there was no apparent difference in the crystal structure. This result indicated that the crystal structure of the material did not change after the photocatalytic reaction.

2.2.2. Effect of Initial Dye Concentration, Initial Solution pH, and the Molar Ratio of Ni^{2+}/Fe^{3+} on the Degradation of RhB

The effect of initial dye concentration on the degradation of RhB over the Ni/Fe-MOF-0.3/Light/H_2O_2 system was evaluated (Figure 10A). As shown in Figure 10A, the degradation efficiency of RhB was slightly decreased when increasing the initial dye concentration from 1×10^{-5} to 4×10^{-5} M. This was mainly because of the increase of the dye molecules around the active sites leading to inhibiting the penetration of light to the surface of the catalyst [66].

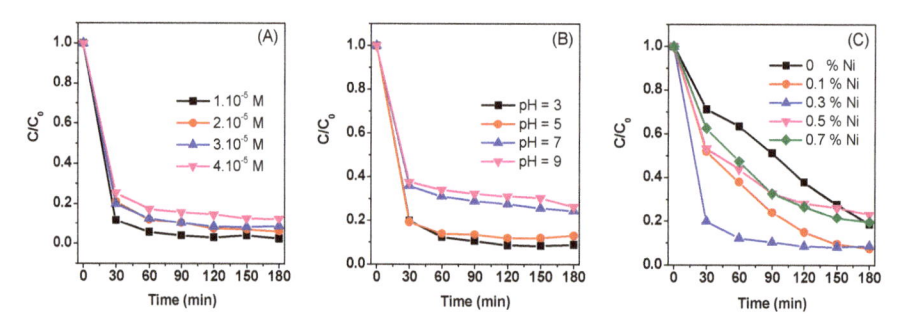

Figure 10. Effect of initial dye concentration (**A**), initial solution pH (**B**), and the molar ratio of Ni^{2+}/Fe^{3+} (**C**) on the degradation of RhB.

The effect of the initial pH on the degradation of RhB on the degradation of RhB over Ni/Fe-MOF/Light/H_2O_2 system was also investigated. The pH of the initial solution was selected as follows: 3, 5 (acidic), 7 (neutral), and 9 (basic). At different pH conditions, the Ni/Fe-MOF-0.3 remained most effective when it came to removing RhB. The RhB removal efficiency peaked at the solution pH of 5 and decreased with increasing pH thereafter (Figure 10B). This result could be explained by the fact that when the pH exceeded the isoelectric point of the material, they were negatively charged. In addition, the RhB used in this experiment was a cationic color such that the material would absorb the color gradually from pH 5 to 9. As the adsorption increased, the color molecules would shield the catalytic surface, which prevented light from irradiating on the catalyst surface, thus decreasing photocatalytic activity and reducing color removal. The pH at the isoelectric point or point of zero charge-pzc of the material was an important parameter for evaluation of the acidity/basicity and the surface charge of the adsorbent in solution. The determination of pHzpc was carried out according to our previously published study [67–69], as follows: Photocatalysts (20 mg) was added to flasks containing 100 mL of KCl 0.1 M at different initial pH values (pH_i = 2, 4, 6, 8, 10, and 12). The solutions were shaken in the shaker for 24 h, and then solids were removed from the mixture by centrifugation at 4000 rpm for 15 min. The final pH of the solution (pH_f) is measured using a pH meter. The curve was plotted via pH_f against the pH_i, and the pHpzc was calculated at $pH_i = pH_f$. As shown in Figure 11A,B, the pH_{pzc} values of the MIL-53(Fe) and Ni/Fe-MOF-0.3 were approximately equal and were within the pH range of 4.1–4.2.

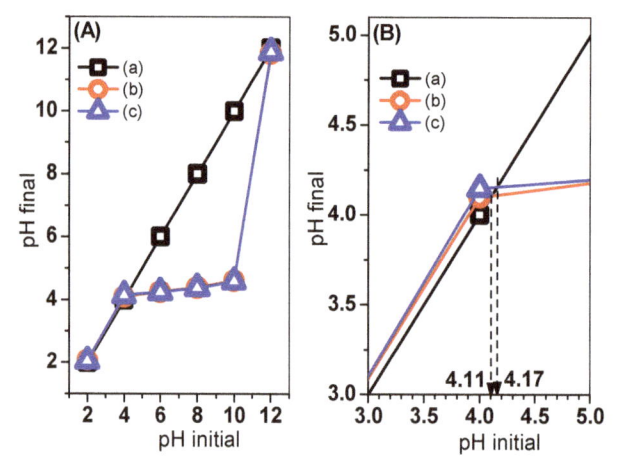

Figure 11. Measurement of pHzpc: the initial versus final pH plot: pH initial (a), pH initial-MIL-53(Fe) (b), and pH initial-Ni/Fe-MOF-0.3 (c) (**A**) and enlarged pH initial from 3 to 5 (**B**).

The degradation results of the different molar ratios of Ni^{2+}/Fe^{3+} in the samples are shown in Figure 10C, where the best performance was obtained with the Ni/Fe-MOF-0.3 sample, followed by the Ni/Fe-MOF-0.1 and Ni/Fe-MOF-0.7 samples. The Ni/Fe-MOF-0.5 sample showed the lowest catalytic activity among all the Ni/Fe-MOF catalysts. This result indicated that the different molar ratio of Ni^{2+}/Fe^{3+} had a significant impact on the photocatalytic performance of Ni/Fe-MOF samples, which may be conducive to the structure and morphology formation of Ni/Fe-MOF.

3. Experimental

3.1. Materials

1,4-Benzenedioic acid (H_2BDC, 98%) and RhB ($\geq$95%) were purchased from Sigma-Aldrich (St. Louis, MO, USA). Iron(III) chloride hexahydrate ($FeCl_3 \cdot 6H_2O$, 99%), nickel(II) nitrate hexahydrate ($Ni(NO_3)_2 \cdot 6H_2O$, 99%), N,N-dimethylformamide (DMF, 99%), ethanol, and hydrogen peroxide (H_2O_2, 30%) were obtained from Xilong Chemical Co., Ltd. (Guangzhou, China). All reagents were used as received without further purification.

3.2. Preparation of Catalysts

Ni/Fe-MOF samples were synthesized using a solvothermal router similar to MIL-53(Fe), according to the previous literature [39]. In a typical synthesis, 9 mmol of H_2BDC, 6 mmol of $FeCl_3 \cdot 6H_2O$, and a certain amount of $Ni(NO_3)_2 \cdot 6H_2O$ were dissolved in 60 mL DMF. The obtained mixture was vigorously stirred for 30 min before being transferred into a 100 mL hydrothermal synthesis autoclave reactor 304 stainless steel high-pressure digestion tank with PTFE lining (Baoshishan Co., Ltd., Shanghai, China). The autoclave was heated at 100 °C in an oven (Memmert UN110, Schwabach, Germany) with a heating rate of 5 °C/min for three days. After being cooled to room temperature in air, the remaining H_2BDC was removed using a distillation method with DMF solvent for 24 h at 100 °C with a heating rate of 5 °C/min. The obtained suspension was centrifuged at 6000 rpm for 30 min, and the orange precipitates located at the bottom of the tube were washed with DMF (three times) and water (three times), respectively. Finally, the product was dried for 24 h at 60 °C. The obtained MOFs samples with corresponding Ni concentration were denoted as Ni/Fe-MOF-x (x is the molar ratio of Ni^{2+}/Fe^{3+}, and was chosen as 0, 0.1, 0.3, 0.5, and 0.7). The specific description is shown in Table S1 and the flow chart of the synthesis method is described in Figure S4. The sample was washed with DMF and water to obtain Ni/Fe-MOF-x·DMF and Ni/Fe-MOF-x·H_2O, respectively. For comparison, MIL(53) also was prepared using a similar method above without the presence of $Ni(NO_3)_2 \cdot 6H_2O$ in the reaction solution mixture.

3.3. Catalyst Characterization

Powder X-ray diffraction (XRD) patterns were conducted on a D8 Advance Bruker powder diffractometer with a Cu Kα source (λ = 0.15405) at a scan rate of 0.04°/s with 2θ = 2 to 30°. The surface morphologies and particle size of Ni/Fe-MOF samples were observed using field emission scanning electron microscope (FESEM, JEOL JSM-7600F, Peabody, MA, USA) equipped with an energy dispersive X-ray spectroscope (EDS, Oxford instruments 50 mm^2 X-Max, Abingdon, UK). FT-IR spectra were recorded on an EQUINOX 55 spectrometer (Bruker, Germany) using the KBr pellet technique. Raman spectroscopy was carried out on the HORIBA Jobin Yvon spectrometer with a laser beam of 633 nm. To examine the existence of Ni and Fe in the samples, X-ray photoelectron spectra (XPS) of the samples was measured using MultiLab 2000 spectrometer (Thermo VG Scientific, Waltham, MA, USA). The optical absorption characteristics of the photocatalysts were determined using ultraviolet-visible (UV/Vis) diffuse reflectance spectroscopy (UV/Vis DRS, Shimazu UV-2450, Kyoto, Japan) in the range 200–900 cm^{-1}. PL spectroscopy was performed using a Hitachi F4500 Fluorescence Spectrometer (Schaumburg, IL, USA) with the Xe Lamp Power range (700–900 V) at room temperature. The specific surface area and pore distribution of MIL-53(Fe) and Ni/Fe-MOFs were determined using the Brunauer–Emmett–Teller (BET) method and Barrett–Joyner–Halenda (BJH) method, respectively (TriStar 3000 V6.07, Micromeritics instrument corporation, Norcross, GA, USA). The samples were kept at 200 °C for 5 h to degas. The pH value was measured using a pH meter (Consort-C1010, Turnhout, Belgium) at room temperature.

3.4. Photocatalytic Test

The photocatalytic activities of Ni/Fe-MOF photocatalysts were evaluated using the photodegradation of RhB under visible light irradiation with a 40 W compact fluorescent lamp (Philips) in the open air and at room temperature (Figure S5). The intensity and wavelength of the light source was 4400 lm and >400 nm, respectively (Figure S6 and Table S2). Therefore, it was suggested that the photocatalytic processes in our experiments were mainly due to the action of the visible light range [70–72]. In each run, a mixture of RhB aqueous solution (3.10^{-5} mol/L, 100 mL), the given catalyst (20 mg), and H_2O_2 (10^{-5} mol/L) was magnetically stirred in the presence or absence of light. Five milliliters of the suspension was withdrawn at the same intervals and immediately centrifuged to separate the photocatalyst particles for 15 min. The concentration of RhB was analyzed using a UV-visible spectrophotometer (Model Evolution 60S, Thermo Fisher Scientific, Waltham, MA, USA) at a maximum absorbance wavelength of $\lambda = 554$ nm. In addition, the effect of parameters including initial dye concentration and initial solution pH on the photodegradation of RhB over Ni/Fe-MOF photocatalysts was also investigated. pH levels of 3, 5, 7, and 9 were selected, whereas the concentrations of RhB were increased from 1.10^{-5} M to 4.10^{-5} M.

4. Conclusions

In summary, we have successfully prepared mixed Ni/Fe-base MOF with different molar ratios of Ni^{2+}/Fe^{3+} via a direct solvothermal approach. The structure characterization results from XRD, Raman, XPS, and FT-IR confirmed that with the presence of mixed ionic Fe^{3+} and Ni^{2+}, MIL-88B crystals based on the mixed metal Fe_2NiO cluster was formed, while MIL-53 (Fe) was formed with the presence of a single ion Fe^{3+}. The photocatalytic performance of the obtained photocatalysts was evaluated in the decolorization of RhB dye. The results indicated that the obtained Ni/Fe-MOF samples exhibited high photocatalytic activity in comparison to MIL-53(Fe). The degradation rate of Ni/Fe-MOF-0.3 could reach the highest (91.14%) after 180 min of visible light irradiation. These results suggest that the Ni/Fe-MOF, which consist mixed-metal cluster Fe_2NiO with electron transfer effects, might enhance the photocatalytic performance.

Supplementary Materials: The following are available online at http://www.mdpi.com/2073-4344/8/11/487/s1, Figure S1: EDS mapping of Ni/Fe-MOF-0.3 sample, Figure S2: UV-vis DRS spectra of as-prepared MIL-53(Fe) and Ni-MIL-53(Fe) crystals isolated from H_2O, Figure S3: XRD patterns of Ni/Fe-MOF-0.3 before and after reactions, Table S1: Synthetic parameters of MIL-53(Fe) and Ni/Fe-MOF samples, Figure S4: The flow chart of the synthesis method, Figure S5: Illustration of the utilized photocatalytic test system, Figure S6. The spectral distribution of a 40 W compact fluorescent lamp, Figure S7: Background-subtracted Fe $2p_{3/2}$ spectrum from Ni/Fe-MOF-0.3, Figure S8: Background-subtracted Fe $2p_{3/2}$ spectrum from MIL-53(Fe), Table S2: Product data of a 40 W compact fluorescent lamp.

Author Contributions: T.D.N. proposed the concept and supervised the research work at Nguyen Tat Thanh University. V.H.N. and Q.T.P.B. designed the experiments and performed the experiments. T.H. and L.D.T. performed XPS and FT-IR analyses. C.V.N. performed SEM and EDS analyses. D.-V.N.V. contributed to the revision of the manuscript. L.G.B. and S.T.D. analyzed the data and wrote the paper.

Funding: This research was funded by NTTU Foundation for Science and Technology Development under grant number 2017.01.13/HĐ-KHCN.

Conflicts of Interest: The authors declare no conflict of interest.

References

1. Tranchemontagne, D.J.; Mendoza-Cortés, J.L.; O'Keeffe, M.; Yaghi, O.M. Secondary building units, nets and bonding in the chemistry of metal-organic frameworks. *Chem. Soc. Rev.* **2009**, *38*, 1257–1283. [CrossRef] [PubMed]
2. Hamon, L.; Serre, C.; Devic, T.; Loiseau, T.; Millange, F.; Férey, G.; De Weireld, G. Comparative study of hydrogen sulfide adsorption in the MIL-53(Al, Cr, Fe), MIL-47(V), MIL-100(Cr), and MIL-101(Cr) metal-organic frameworks at room temperature. *J. Am. Chem. Soc.* **2009**, *131*, 8775–8777. [CrossRef] [PubMed]
3. Devic, T.; Salles, F.; Bourrelly, S.; Moulin, B.; Maurin, G.; Horcajada, P.; Serre, C.; Vimont, A.; Lavalley, J.C.; Leclerc, H.; et al. Effect of the organic functionalization of flexible MOFs on the adsorption of CO_2. *J. Mater. Chem.* **2012**, *22*, 10266–10273. [CrossRef]
4. Vu, T.A.; Le, G.H.; Dao, C.D.; Dang, L.Q.; Nguyen, K.T.; Nguyen, Q.K.; Dang, P.T.; Tran, H.T.K.; Duong, Q.T.; Nguyen, T.V.; et al. Arsenic removal from aqueous solutions by adsorption using novel MIL-53(Fe) as a highly efficient adsorbent. *RSC Adv.* **2015**, *5*, 5261–5268. [CrossRef]
5. Jia, J.; Xu, F.; Long, Z.; Hou, X.; Sepaniak, M.J. Metal–organic framework MIL-53(Fe) for highly selective and ultrasensitive direct sensing of MeHg+. *Chem. Commun.* **2013**, *49*, 4670. [CrossRef] [PubMed]
6. Gao, X.; Zhai, M.; Guan, W.; Liu, J.; Liu, Z.; Damirin, A. Controllable synthesis of a smart multifunctional nanoscale metal-organic framework for magnetic resonance/optical imaging and targeted drug delivery. *ACS Appl. Mater. Interfaces* **2017**, *9*, 3455–3462. [CrossRef] [PubMed]
7. Bach, L.G.; Van Tran, T.; Nguyen, T.D.; Van Pham, T.; Do, S.T. Enhanced adsorption of methylene blue onto graphene oxide-doped XFe_2O_4(X = Co, Mn, Ni) nanocomposites: Kinetic, isothermal, thermodynamic and recyclability studies. *Res. Chem. Intermed.* **2018**, *44*, 1661–1687. [CrossRef]
8. Nhung, N.T.H.; Quynh, B.T.P.; Thao, P.T.T.; Bich, H.N.; Giang, B.L. Pretreated Fruit Peels as Adsorbents for Removal of Dyes from Water. *IOP Conf. Ser. Earth Environ. Sci.* **2018**, *159*, 012015. [CrossRef]
9. Kim, D.W.; Bach, L.G.; Hong, S.S.; Park, C.; Lim, K.T. A Facile Route towards the Synthesis of Fe_3O_4/Graphene Oxide Nanocomposites for Environmental Applications. *Mol. Cryst. Liq. Cryst.* **2014**. [CrossRef]
10. Van Thuan, T.; Ho, V.T.T.; Trinh, N.D.; Thuong, N.T.; Quynh, B.T.P.; Bach, L.G. Facile one-spot synthesis of highly porous KOH-activated carbon from rice husk: Response surface methodology approach. *Carbon Sci. Technol.* **2016**, *8*, 63–69.
11. Trinh, N.D.; Hong, S.-S. Photocatalytic Decomposition of Methylene Blue Over MIL-53 (Fe) Prepared Using Microwave-Assisted Process Under Visible Light Irradiation. *J. Nanosci. Nanotechnol.* **2015**, *15*, 5450–5454. [CrossRef] [PubMed]
12. Yang, S.J.; Im, J.H.; Kim, T.; Lee, K.; Park, C.R. MOF-derived ZnO and ZnO@C composites with high photocatalytic activity and adsorption capacity. *J. Hazard. Mater.* **2011**, *186*, 376–382. [CrossRef] [PubMed]
13. Du, J.J.; Yuan, Y.P.; Sun, J.X.; Peng, F.M.; Jiang, X.; Qiu, L.G.; Xie, A.J.; Shen, Y.H.; Zhu, J.F. New photocatalysts based on MIL-53 metal-organic frameworks for the decolorization of methylene blue dye. *J. Hazard. Mater.* **2011**, *190*, 945–951. [CrossRef] [PubMed]
14. Zhang, C.; Ai, L.; Jiang, J. Solvothermal synthesis of MIL-53(Fe) hybrid magnetic composites for photoelectrochemical water oxidation and organic pollutant photodegradation under visible light. *J. Mater. Chem. A* **2015**, *3*, 3074–3081. [CrossRef]
15. Alhamami, M.; Doan, H.; Cheng, C.H. A review on breathing behaviors of metal-organic-frameworks (MOFs) for gas adsorption. *Materials* **2014**, *7*, 3198–3250. [CrossRef] [PubMed]
16. Dhakshinamoorthy, A.; Alvaro, M.; Garcia, H. Commercial metal-organic frameworks as heterogeneous catalysts. *Chem. Commun.* **2012**, *48*, 11275–11288. [CrossRef] [PubMed]
17. Ai, L.; Zhang, C.; Li, L.; Jiang, J. Iron terephthalate metal-organic framework: Revealing the effective activation of hydrogen peroxide for the degradation of organic dye under visible light irradiation. *Appl. Catal. B Environ.* **2014**, *148–149*, 191–200. [CrossRef]
18. Lionet, Z.; Kamata, Y.; Nishijima, S.; Toyao, T.; Kim, T.-H.; Horiuchi, Y.; Lee, S.W.; Matsuoka, M. Water oxidation reaction promoted by MIL-101(Fe) photoanode under visible light irradiation. *Res. Chem. Intermed.* **2018**, *44*, 4755–4764. [CrossRef]
19. Qu, L.-L.; Wang, J.; Xu, T.-Y.; Chen, Q.-Y.; Chen, J.-H.; Shi, C.-J. Iron(iii)-based metal–organic frameworks as oxygen-evolving photocatalysts for water oxidation. *Sustain. Energy Fuels* **2018**, *2*, 2109–2114. [CrossRef]

20. Wang, D.; Li, Z. Iron-based metal–organic frameworks (MOFs) for visible-light-induced photocatalysis. *Res. Chem. Intermed.* **2017**, *43*, 5169–5186. [CrossRef]
21. Wang, D.; Albero, J.; García, H.; Li, Z. Visible-light-induced tandem reaction of o-aminothiophenols and alcohols to benzothiazoles over Fe-based MOFs: Influence of the structure elucidated by transient absorption spectroscopy. *J. Catal.* **2017**, *349*, 156–162. [CrossRef]
22. Horiuchi, Y.; Toyao, T.; Miyahara, K.; Zakary, L.; Do Van, D.; Kamata, Y.; Kim, T.-H.; Lee, S.W.; Matsuoka, M. Visible-light-driven photocatalytic water oxidation catalysed by iron-based metal–organic frameworks. *Chem. Commun.* **2016**, *52*, 5190–5193. [CrossRef] [PubMed]
23. Hu, L.; Deng, G.; Lu, W.; Pang, S.; Hu, X. Deposition of CdS nanoparticles on MIL-53(Fe) metal-organic framework with enhanced photocatalytic degradation of RhB under visible light irradiation. *Appl. Surf. Sci.* **2017**, *410*, 401–413. [CrossRef]
24. Zhou, Y.; Mao, Z.; Wang, W.; Yang, Z.; Liu, X. In-Situ Fabrication of Graphene Oxide Hybrid Ni-Based Metal-Organic Framework (Ni-MOFs@GO) with Ultrahigh Capacitance as Electrochemical Pseudocapacitor Materials. *ACS Appl. Mater. Interfaces* **2016**, *8*, 28904–28916. [CrossRef] [PubMed]
25. Panda, R.; Rahut, S.; Basu, J.K. Preparation of a Fe_2O_3/MIL-53(Fe) composite by partial thermal decomposition of MIL-53(Fe) nanorods and their photocatalytic activity. *RSC Adv.* **2016**, *6*, 80981–80985. [CrossRef]
26. Lou, X.; Hu, H.; Li, C.; Hu, X.; Li, T.; Shen, M.; Chen, Q.; Hu, B. Capacity control of ferric coordination polymers by zinc nitrate for lithium-ion batteries. *RSC Adv.* **2016**, *6*, 86126–86130. [CrossRef]
27. Vuong, G.T.; Pham, M.H.; Do, T.O. Synthesis and engineering porosity of a mixed metal Fe2Ni MIL-88B metal-organic framework. *Dalt. Trans.* **2013**, *42*, 550–557. [CrossRef] [PubMed]
28. Vuong, G.-T.; Pham, M.-H.; Do, T.-O. Direct synthesis and mechanism of the formation of mixed metal Fe2Ni-MIL-88B. *CrystEngComm* **2013**, *15*, 9694. [CrossRef]
29. Pham, M.-H.; Dinh, C.-T.; Vuong, G.-T.; Ta, N.-D.; Do, T.-O. Visible light induced hydrogen generation using a hollow photocatalyst with two cocatalysts separated on two surface sides. *Phys. Chem. Chem. Phys.* **2014**, *16*, 5937. [CrossRef] [PubMed]
30. Sun, Q.; Liu, M.; Li, K.; Han, Y.; Zuo, Y.; Chai, F.; Song, C.; Zhang, G.; Guo, X. Synthesis of Fe/M (M = Mn, Co, Ni) bimetallic metal organic frameworks and their catalytic activity for phenol degradation under mild conditions. *Inorg. Chem. Front.* **2017**, *4*, 144–153. [CrossRef]
31. Díaz-Gallifa, P.; Fabelo, O.; Pasán, J.; Cañadillas-Delgado, L.; Lloret, F.; Julve, M.; Ruiz-Pérez, C. Two-dimensional 3d-4f heterometallic coordination polymers: Syntheses, crystal structures, and magnetic properties of six new Co(II)-Ln(III) compounds. *Inorg. Chem.* **2014**, *53*, 6299–6308. [CrossRef] [PubMed]
32. Vu, T.A.; Le, G.H.; Dao, C.D.; Dang, L.Q.; Nguyen, K.T.; Dang, P.T.; Tran, H.T.K.; Duong, Q.T.; Nguyen, T.V.; Lee, G.D. Isomorphous substitution of Cr by Fe in MIL-101 framework and its application as a novel heterogeneous photo-Fenton catalyst for reactive dye degradation. *RSC Adv.* **2014**, *40*, 41185–41194. [CrossRef]
33. Zhu, W.; Liu, P.; Xiao, S.; Wang, W.; Zhang, D.; Li, H. Microwave-assisted synthesis of Ag-doped MOFs-like organotitanium polymer with high activity in visible-light driven photocatalytic NO oxidization. *Appl. Catal. B Environ.* **2015**, *172–173*, 46–51. [CrossRef]
34. Han, Y.; Liu, M.; Li, K.; Sun, Q.; Zhang, W.; Song, C.; Zhang, G.; Conrad Zhang, Z.; Guo, X. In situ synthesis of titanium doped hybrid metal-organic framework UiO-66 with enhanced adsorption capacity for organic dyes. *Inorg. Chem. Front.* **2017**, *4*, 1870–1880. [CrossRef]
35. Vinh, N.H.; Long Giang, B.; Sy Trung, D.; Thuong, N.T.; Trinh, N.D. Photoluminescence Properties of Eu-Doped MIL-53(Fe) Obtained by Solvothermal Synthesis. *J. Nanosci. Nanotechnol.* **2019**, *19*, 1148–1150.
36. Chi, Y.X.; Niu, S.Y.; Jin, J. Syntheses, structures and photophysical properties of a series of Zn-Ln coordination polymers (Ln = Nd, Pr, Sm, Eu, Tb, Dy). *Inorg. Chim. Acta* **2009**, *362*, 3821–3828. [CrossRef]
37. Tahir, M. Synergistic effect in MMT-dispersed Au/TiO_2 monolithic nanocatalyst for plasmon-absorption and metallic interband transitions dynamic CO_2 photo-reduction to CO. *Appl. Catal. B Environ.* **2017**, *219*, 329–343. [CrossRef]
38. Tahir, M. Ni/MMT-promoted TiO_2 nanocatalyst for dynamic photocatalytic H_2 and hydrocarbons production from ethanol-water mixture under UV-light. *Int. J. Hydrogen Energy* **2017**, *42*, 28309–28326. [CrossRef]
39. Haque, E.; Khan, N.; Park, H.J.; Jhung, S.H. Synthesis of a metal-organic framework material, iron terephthalate, by ultrasound, microwave, and conventional electric heating: A kinetic study. *Chem. A Eur. J.* **2010**, *16*, 1046–1052. [CrossRef] [PubMed]

40. Pu, M.; Guan, Z.; Ma, Y.; Wan, J.; Wang, Y.; Brusseau, M.L.; Chi, H. Synthesis of iron-based metal-organic framework MIL-53 as an efficient catalyst to activate persulfate for the degradation of Orange G in aqueous solution. *Appl. Catal. A Gen.* **2018**, *549*, 82–92. [CrossRef] [PubMed]
41. Feng, X.; Chen, H.; Jiang, F. In-situ ethylenediamine-assisted synthesis of a magnetic iron-based metal-organic framework MIL-53(Fe) for visible light photocatalysis. *J. Colloid Interface Sci.* **2017**, *494*, 32–37. [CrossRef] [PubMed]
42. Shannon, R.D. Revised effective ionic radii and systematic studies of interatomic distances in halides and chalcogenides. *Acta Crystallogr. Sect. A* **1976**, *32*, 751–767. [CrossRef]
43. Nakamoto, K. Applications in Organometallic Chemistry. In *Infrared and Raman Spectra of Inorganic and Coordination Compounds*; John Wiley & Sons, Inc.: Hoboken, NJ, USA; pp. 275–331, ISBN 9780470405888.
44. Gao, Y.; Li, S.; Li, Y.; Yao, L.; Zhang, H. Accelerated photocatalytic degradation of organic pollutant over metal-organic framework MIL-53(Fe) under visible LED light mediated by persulfate. *Appl. Catal. B Environ.* **2017**, *202*, 165–174. [CrossRef]
45. Vu, T.A.; Le, G.H.; Vu, H.T.; Nguyen, K.T.; Quan, T.T.T.; Nguyen, Q.K.; Tran, H.T.K.; Dang, P.T.; Vu, L.D.; Lee, G.D. Highly photocatalytic activity of novel Fe-MIL-88B/GO nanocomposite in the degradation of reactive dye from aqueous solution. *Mater. Res. Express* **2017**, *4*, 035038. [CrossRef]
46. Islam, M.R.; Bach, L.G.; Vo, T.S.; Tran, T.N.; Lim, K.T. Nondestructive chemical functionalization of MWNTs by poly(2-dimethylaminoethyl methacrylate) and their conjugation with CdSe quantum dots: Synthesis, properties, and cytotoxicity studies. *Appl. Surf. Sci.* **2013**. [CrossRef]
47. Grosvenor, A.P.; Kobe, B.A.; Biesinger, M.C.; McIntyre, N.S. Investigation of multiplet splitting of Fe 2p XPS spectra and bonding in iron compounds. *Surf. Interface Anal.* **2004**, *36*, 1564–1574. [CrossRef]
48. Molchan, I.S.; Thompson, G.E.; Skeldon, P.; Lindsay, R.; Walton, J.; Kouvelos, E.; Romanos, G.E.; Falaras, P.; Kontos, A.G.; Arfanis, M.; et al. Microscopic study of the corrosion behaviour of mild steel in ionic liquids for CO_2 capture applications. *RSC Adv.* **2015**, *5*, 35181–35194. [CrossRef]
49. Lian, K.K.; Kirk, D.W.; Thorpe, S.J. Investigation of a "Two-State" Tafel Phenomenon for the Oxygen Evolution Reaction on an Amorphous Ni-Co Alloy. *J. Electrochem. Soc.* **1995**, *142*, 3704–3712. [CrossRef]
50. Siang, T.J.; Bach, L.G.; Singh, S.; Truong, Q.D.; Ho, V.T.T.; Huy Phuc, N.H.; Alenazey, F.; Vo, D.V.N. Methane bi-reforming over boron-doped Ni/SBA-15 catalyst: Longevity evaluation. *Int. J. Hydrogen Energy* **2018**. [CrossRef]
51. Sing, K.S.W. Reporting physisorption data for gas/solid systems with special reference to the determination of surface area and porosity (Recommendations 1984). *Pure Appl. Chem.* **1985**, *57*, 603–619. [CrossRef]
52. Horcajada, P.; Serre, C.; Maurin, G.; Ramsahye, N.A.; Balas, F.; Vallet-Regí, M.; Sebban, M.; Taulelle, F.; Férey, G. Flexible porous metal-organic frameworks for a controlled drug delivery. *J. Am. Chem. Soc.* **2008**, *130*, 6774–6780. [CrossRef] [PubMed]
53. Blake, A.B.; Yavari, A.; Hatfield, W.E.; Sethulekshmi, C.N. Magnetic and spectroscopic properties of some heterotrinuclear basic acetates of chromium(III), iron(III), and divalent metal ions. *J. Chem. Soc. Dalt. Trans.* **1985**, 2509–2520. [CrossRef]
54. Zhang, Z.; Li, X.; Liu, B.; Zhao, Q.; Chen, G. Hexagonal microspindle of NH2-MIL-101(Fe) metal-organic frameworks with visible-light-induced photocatalytic activity for the degradation of toluene. *RSC Adv.* **2016**, *6*, 4289–4295. [CrossRef]
55. Nguyen, M.T.H.; Nguyen, Q.T. Efficient refinement of a metal-organic framework MIL-53(Fe) by UV-vis irradiation in aqueous hydrogen peroxide solution. *J. Photochem. Photobiol. A Chem.* **2014**, *288*, 55–59. [CrossRef]
56. Ling, S.; Slater, B. Unusually Large Band Gap Changes in Breathing Metal-Organic Framework Materials. *J. Phys. Chem. C* **2015**, *119*, 16667–16677. [CrossRef]
57. Hou, Y.; Yuan, H.; Chen, H.; Feng, J.; Ding, Y.; Li, L. Preparation of La^{3+}/Zn^{2+}-doped BiVO4 nanoparticles and its enhanced visible photocatalytic activity. *Appl. Phys. A* **2017**, *123*, 611. [CrossRef]
58. Dutta, D.P.; Ballal, A.; Chopade, S.; Kumar, A. A study on the effect of transition metal (Ti^{4+}, Mn^{2+}, Cu^{2+} and Zn^{2+})-doping on visible light photocatalytic activity of Bi_2MoO_6 nanorods. *J. Photochem. Photobiol. A Chem.* **2017**, *346*, 105–112. [CrossRef]
59. Jin, A.Z.; Dong, W.; Yang, M.; Wang, J.; Wang, G.; Jin, Z.; Dong, W.; Yang, M.; Wang, J.; Wang, G. One-pot Preparation of Hierarchical Nanosheet-Constructed Fe_3O_4/MIL-88B (Fe) Magnetic Microspheres with High Efficiency Photocatalytic Degradation of Dye. *ChemCatChem* **2016**, *22*, 3510–3517. [CrossRef]

60. Liang, R.; Luo, S.; Jing, F.; Shen, L.; Qin, N.; Wu, L. A simple strategy for fabrication of Pd@MIL-100(Fe) nanocomposite as a visible-light-driven photocatalyst for the treatment of pharmaceuticals and personal care products (PPCPs). *Appl. Catal. B Environ.* **2015**, *176–177*, 240–248. [CrossRef]
61. Laurier, K.G.M.; Vermoortele, F.; Ameloot, R.; De Vos, D.E.; Hofkens, J.; Roeffaers, M.B.J. Iron(III)-based metal-organic frameworks as visible light photocatalysts. *J. Am. Chem. Soc.* **2013**, *135*, 14488–14491. [CrossRef] [PubMed]
62. Martínez, F.; Leo, P.; Orcajo, G.; Díaz-García, M.; Sanchez-Sanchez, M.; Calleja, G. Sustainable Fe-BTC catalyst for efficient removal of mehylene blue by advanced fenton oxidation. *Catal. Today* **2018**, *313*, 6–11. [CrossRef]
63. Gao, C.; Chen, S.; Quan, X.; Yu, H.; Zhang, Y. Enhanced Fenton-like catalysis by iron-based metal organic frameworks for degradation of organic pollutants. *J. Catal.* **2017**, *356*, 125–132. [CrossRef]
64. Liu, Q.; Liu, Y.; Gao, B.; Chen, Y.; Lin, B. Hydrothermal synthesis of In_2O_3-loaded $BiVO_4$ with exposed {010}{110} facets for enhanced visible-light photocatalytic activity. *Mater. Res. Bull.* **2017**, *87*, 114–118. [CrossRef]
65. Sotnik, S.A.; Polunin, R.A.; Kiskin, M.A.; Kirillov, A.M.; Dorofeeva, V.N.; Gavrilenko, K.S.; Eremenko, I.L.; Novotortsev, V.M.; Kolotilov, S.V. Heterometallic coordination polymers assembled from trigonal trinuclear Fe_2Ni-pivalate blocks and polypyridine spacers: Topological diversity, sorption, and catalytic properties. *Inorg. Chem.* **2015**, *54*, 5169–5181. [CrossRef] [PubMed]
66. Jiang, Q.; Håkansson, M.; Suomi, J.; Ala-Kleme, T.; Kulmala, S. Cathodic electrochemiluminescence of lucigenin at disposable oxide-coated aluminum electrodes. *J. Electroanal. Chem.* **2006**, *591*, 85–92. [CrossRef]
67. Van Thuan, T.; Quynh, B.T.P.; Nguyen, T.D.; Ho, V.T.T.; Bach, L.G. Response surface methodology approach for optimization of Cu^{2+}, Ni^{2+} and Pb^{2+} adsorption using KOH-activated carbon from banana peel. *Surf. Interfaces* **2017**, *6*, 209–217. [CrossRef]
68. Van Tran, T.; Bui, Q.T.P.; Nguyen, T.D.; Le, N.T.H.; Bach, L.G. A comparative study on the removal efficiency of metal ions (Cu^{2+}, Ni^{2+}, and Pb^{2+}) using sugarcane bagasse-derived ZnCl2-activated carbon by the response surface methodology. *Adsorpt. Sci. Technol.* **2017**, *35*, 72–85. [CrossRef]
69. Van Tran, T.; Bui, Q.T.P.; Nguyen, T.D.; Thanh Ho, V.T.; Bach, L.G. Application of response surface methodology to optimize the fabrication of $ZnCl_2$-activated carbon from sugarcane bagasse for the removal of Cu^{2+}. *Water Sci. Technol.* **2017**, *75*, 2047–2055. [CrossRef] [PubMed]
70. Dinh, C.-T.; Nguyen, T.-D.; Kleitz, F.; Do, T.-O. Large-scale synthesis of uniform silver orthophosphate colloidal nanocrystals exhibiting high visible light photocatalytic activity. *Chem. Commun.* **2011**, *47*, 7797–7799. [CrossRef] [PubMed]
71. Yu, X.; Cohen, S.M. Photocatalytic metal-organic frameworks for the aerobic oxidation of arylboronic acids. *Chem. Commun.* **2015**, *51*, 9880–9883. [CrossRef] [PubMed]
72. Lam, S.M.; Sin, J.C.; Abdullah, A.Z.; Mohamed, A.R. Efficient photodegradation of resorcinol with Ag_2O/ZnO nanorods heterostructure under a compact fluorescent lamp irradiation. *Chem. Pap.* **2013**, *67*, 1277–1284. [CrossRef]

© 2018 by the authors. Licensee MDPI, Basel, Switzerland. This article is an open access article distributed under the terms and conditions of the Creative Commons Attribution (CC BY) license (http://creativecommons.org/licenses/by/4.0/).

Article

Photocatalytic Behavior of Strontium Aluminates Co-Doped with Europium and Dysprosium Synthesized by Hydrothermal Reaction in Degradation of Methylene Blue

Byung-Geon Park

Department of Food and Nutrition, Kwangju Women's University, 165 Sanjung-dong, Gwangju 62396, Korea; bgpark@kwu.ac.kr or bgpark814@daum.net; Tel.: +82-62-950-0814

Received: 4 April 2018; Accepted: 22 May 2018; Published: 28 May 2018

Abstract: Strontium aluminates co-doped with europium and dysprosium were prepared by a hydrothermal reaction through a sintering process at lower temperatures. The physicochemical properties of the strontium aluminates co-doped with europium and dysprosium were characterized and compared with those of strontium aluminates prepared by a sol–gel method. The photocatalytic properties of the strontium aluminates co-doped with europium and dysprosium were evaluated through the photocatalytic decomposition of methylene blue dye. The strontium aluminates co-doped with europium and dysprosium prepared by the hydrothermal reaction exhibited good phosphorescence and photocatalytic activities that were similar to those prepared by the sol–gel method. The photocatalytic activity of these catalysts for methylene blue degradation was higher than that of the titanium dioxide (TiO_2) photocatalyst.

Keywords: strontium aluminates; dye photodecomposition; hydrothermal reaction; sol–gel method; phosphorescence

1. Introduction

Alkaline earth aluminates have attracted considerable attention as long afterglow materials because of their excellent photoluminescence, radiation intensity, color purity, and good radiation resistance [1]. In particular, strontium aluminates (SAO) co-doped with europium and dysprosium (SAO; $SrAl_2O_4$: Eu^{2+}, Dy^{3+}) are used in many fields owing to their excellent phosphorescence [2]. $SrAl_2O_4$: Eu^{2+}, Dy^{3+} is applied in emergency lighting, safe indications, signposts, graphic art, billboards, and interior design [3–5]. In addition, the material can be used to synthesize new metal compound composites [6] as well as cathoderay tubes and plasma display panels [7,8]. They also exhibit photocatalytic activity owing to their photosensitive properties [9].

The sol–gel process has attracted considerable interest in obtaining novel chemical compositions and relatively lower reaction temperatures, resulting in homogeneous products [9]. The process enables the synthesis of phosphors with a small size. The inorganic salt-based sol–gel approach has attracted greater interest than the alkoxide-based sol–gel process in the preparation of strontium aluminate luminescent materials [10,11] because inorganic salts are usually non-toxic and cheaper than alkoxides.

Many methods for preparing SAOs have been reported, such as high temperature solid-state reactions [12,13], sol–gel methods [14–16], co-precipitation methods [17], and hydrothermal reaction methods [18,19]. To prepare SAOs by sol–gel method, the mixed reactant sol process should be calcined at temperatures higher than 1000 °C as lower temperatures will not lead to SAOs with good crystallinity [20]. The calcination temperature should be lowered to accomplish low-cost preparation of SAOs.

This paper reports the preparation of SAOs by a hydrothermal reaction through a sintering process at lower temperatures. The physical properties and phosphorescence of SAOs were characterized and compared with those of SAOs prepared by the sol–gel method. The photocatalytic decomposition of methylene blue (MB) dye using SAO was performed to estimate its photocatalytic activities for the photocatalytic degradation of methylene blue dye.

2. Results and Discussion

2.1. Physicochemical Properties of the SAOs

Figure 1 presents X-ray diffraction (XRD) patterns of the SAOs obtained by a hydrothermal reaction and sol–gel method. Single-phase SAOs were obtained from the two methods. The positions and intensities of the main peaks of the two SAOs corresponded entirely to the standard card (No 34-0379). This suggests that the products were the $SrAl_2O_4$ phase. The XRD patterns of SAOs prepared by hydrothermal reaction showed many reflections. Small quantities of $Sr_3Al_2O_6$ and $SrAl_4O_7$ were observed on the particle surface. Figure 2 presents scanning electron microscopy (SEM) images of the SAOs synthesized by the hydrothermal reaction and sol–gel methods. The SAOs were polycrystalline, and the particles were sintered into irregular shapes due to the high calcination temperature. The crystallinity of SAOs prepared by the sol–gel method was superior to that prepared by the hydrothermal reaction. In particular, SEM images of the SAOs prepared by a hydrothermal reaction showed smaller crystals on the surface of the particles. This suggests that the smaller crystals on the particle surface in the SEM images are the $Sr_3Al_2O_6$ and $SrAl_4O_7$ phases.

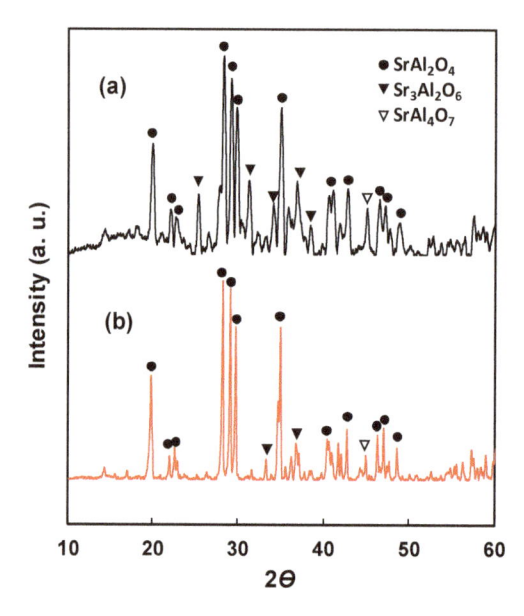

Figure 1. XRD patterns of strontium aluminates (SAO) synthesized by (a) hydrothermal reaction and (b) sol–gel method.

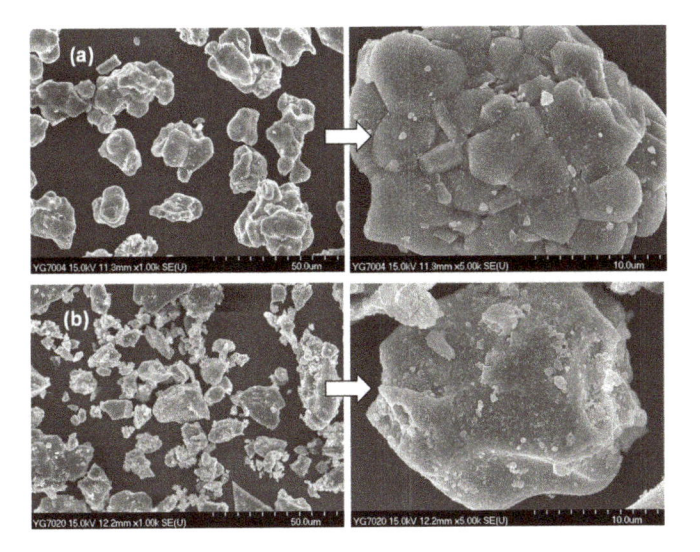

Figure 2. Scanning electron microscopy (SEM) images of SAOs synthesized by (**a**) hydrothermal reaction and (**b**) sol–gel method.

Figure 3 presents the energy-dispersive X-ray (EDX) spectra of the SAOs prepared by the hydrothermal reaction and sol–gel methods. Peaks for Sr and Al were observed. The intensities of the Sr and Al peaks in SAOs prepared by a hydrothermal reaction were similar to those prepared by the sol–gel method. Figure 4 presents Fourier transform infrared (FTIR) spectra of the SAOs. $SrAl_2O_4$ belongs to a distorted stuffed tridymite structure. Tridymite is a member of the nepheline family of structures consisting of a corner-sharing tetrahedral framework that distorts to form large cation-occupying cavities. In $SrAl_2O_4$, the framework is built up by AlO_4 tetrahedra and the structural channels are occupied by Sr^{2+} ions [21]. The XO_4 molecule will have four degenerate normal modes of vibrations: Symmetric stretching (γ_s), symmetric bending (δ_s), antisymmetric stretching (γ_{as}), and antisymmetric bending (δ_d) [22].

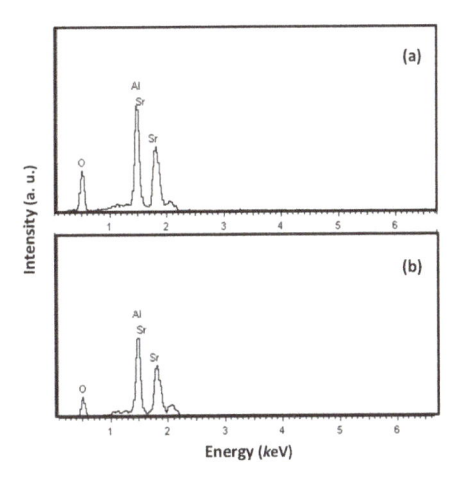

Figure 3. Energy-dispersive X-ray (EDX) spectra of SAOs synthesized by (**a**) hydrothermal reaction and (**b**) sol–gel method.

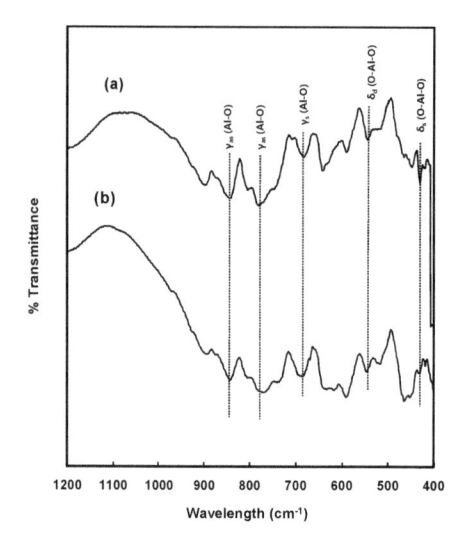

Figure 4. Infrared spectra of the SAOs prepared by (**a**) hydrothermal reaction and (**b**) sol–gel method.

Figure 5 presents the N_2 isotherm of SAOs prepared by a hydrothermal reaction. SAOs are composed of single crystals, as defined in the SEM images. The type of isotherm of SAOs indicated the typical adsorption pattern of nonporous particles. The hysteresis in the isotherm curve was derived from some crevices between the particles. The specific surface areas of the SAOs determined from the Brunauer–Emmett–Teller (BET) equation were 62.5 m^2/g and 51.6 m^2/g, respectively.

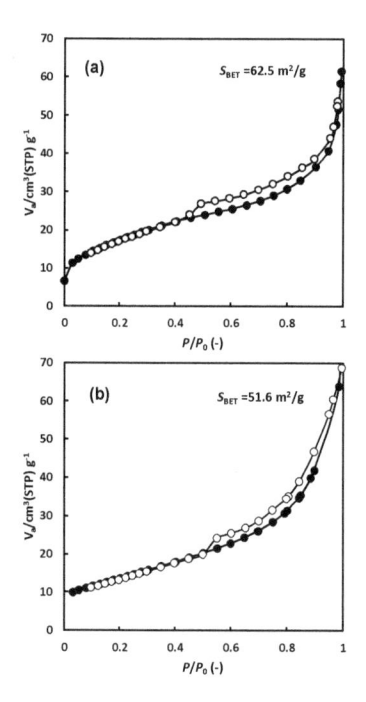

Figure 5. N_2 isotherm of SAOs prepared by (**a**) hydrothermal reaction and (**b**) sol–gel method.

2.2. Luminescent Properties of the SAO Products

Figure 6 presents the emission spectra of the SAOs prepared by the hydrothermal reaction and sol–gel methods. The luminescence properties of the SAO particles were measured in the solid state at room temperature. Regardless of the preparation methods, the SAO particles exhibited similar emission peaks centered at approximately 615 nm under an excitation of 266 nm. The intensities of the peaks for the SAOs prepared by both synthesis methods were similar, which was associated with the typical $4f^65d^1 \rightarrow 4f^7$ transition of the Eu^{2+} ion in $SrAl_2O_4$. This strongly affected the nature of the Eu^{2+} surroundings, where the shielding function of the electrons in the inner shell split the mixed states of 4f and 5d by the crystal field [23]. The special emissions of Dy^{3+} and Eu^{3+} were not observed in the spectra. The Eu^{3+} ions in the precursor were reduced to Eu^{2+} in a weak reducing atmosphere. The Eu^{2+} ions in the precursor were reduced to Eu^+. The Dy^{3+} ions were oxidized to Dy^{4+} during excitation [24]. Simultaneously, thermal vibrations of the surrounding ions and local vibrations in the lattice structure resulted in luminescence spectra with broad bands [25]. The SAOs prepared by the hydrothermal reaction through lower temperature calcination exhibited a similar emission intensity to the SAOs prepared by the sol–gel method through a higher temperature calcination.

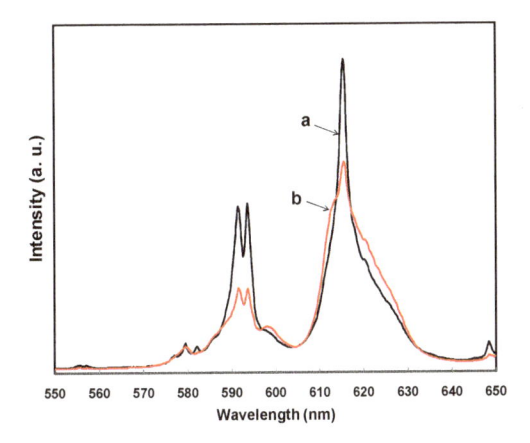

Figure 6. Emission spectra of the SAOs prepared by (a) hydrothermal reaction and (b) sol–gel method.

Figure 7 presents the UV-visible diffuse reflectance spectra (DRS) results of the SAOs and titanium dioxide (TiO_2) converted to Kubelka–Munk units. The optical bandgap (E_{gap}) was calculated using the method proposed by Kubelka and Munk for indirect electronic transitions [26]. The Kubelka–Munk equation is expressed as $F(R) = (1 - R)^2/2R = K/S$, where R, K, and S are the absolute reflectance, absorption coefficient, and scattering coefficient, respectively. The optical properties of the SAOs were induced by light absorption in the photochemical reaction. The diffuse reflectance spectrum of TiO_2 exhibited an adsorption edge at ca. 380 nm. The bandgap of TiO_2 determined from the adsorption edge was 3.2 eV. By contrast, the DRS of the SAOs were shifted to the upper wavelength range. The SAOs exhibited a significant increase in wavelength. The bandgap of the SAOs was ca. 2.9 eV.

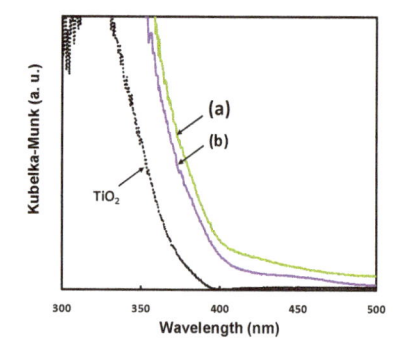

Figure 7. UV-visible diffuse reflectance spectra (DRS) of the SAOs prepared by (**a**) hydrothermal reaction and (**b**) sol–gel method.

2.3. Photocatalytic Properties of the SAO

Figure 8A presents the UV-vis spectra of a pure MB solution, MB solution containing SAO prepared by a hydrothermal reaction, and MB solution containing SAO prepared by the sol–gel method. The absorbance increased with the injection of SAOs prepared by different methods in the MB solution. Figure 8B presents the absorbance of the MB solution containing SAO prepared by hydrothermal reaction and the sol–gel method as a function of the irradiation time of UV light at 300 nm. The absorbance decreased with the injection of SAO into the MB solution despite the short irradiation time. This indicates that the injection of SAOs into the MB solution leads to rapid degradation of the MB dye, and the degradation rate was higher in the MB solution containing SAO prepared by the sol–gel method.

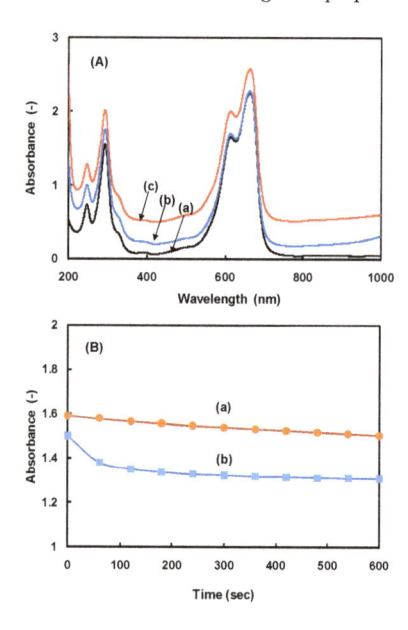

Figure 8. Results of (**A**) UV-vis spectra of as a function of wavelength of (**a**) methylene blue(MB) solution, (**b**) MB solution containing SAO prepared by hydrothermal reaction, and (**c**) MB solution containing SAO prepared by sol–gel method; (**B**) variation of absorbance of the MB solution containing (**a**) SAO prepared by hydrothermal reaction, and (**b**) SAO prepared by sol–gel method as a function of irradiation time of UV light at 300 nm.

Figure 9a presents the changes in MB concentration with different initial MB concentrations on a TiO_2 photocatalyst. The MB concentrations decreased due to the photocatalytic decomposition of MB. The rate of MB degradation was faster at lower initial concentrations of MB than at higher initial concentrations because the photoefficiency increases with decreasing dye concentration. In addition, a large amount of dye might be adsorbed on the TiO_2 surface, which can prevent the dye molecules from coming in contact with the free radicals and electron holes. Figure 9b shows the photocatalytic degradation of MB on TiO_2 and the SAO photocatalysts at the same initial concentration of MB. The degradation of MB was faster with the SAO photocatalysts than that with the TiO_2 photocatalyst. This suggests that SAOs have higher photocatalytic activity than the TiO_2 photocatalyst. The higher photocatalytic activity of SAO was attributed to its higher photosensitivity, which was defined in the DRS results. The SAOs showed a lower bandgap than TiO_2. A lower bandgap of SAOs led to enhanced photocatalytic activity than TiO_2.

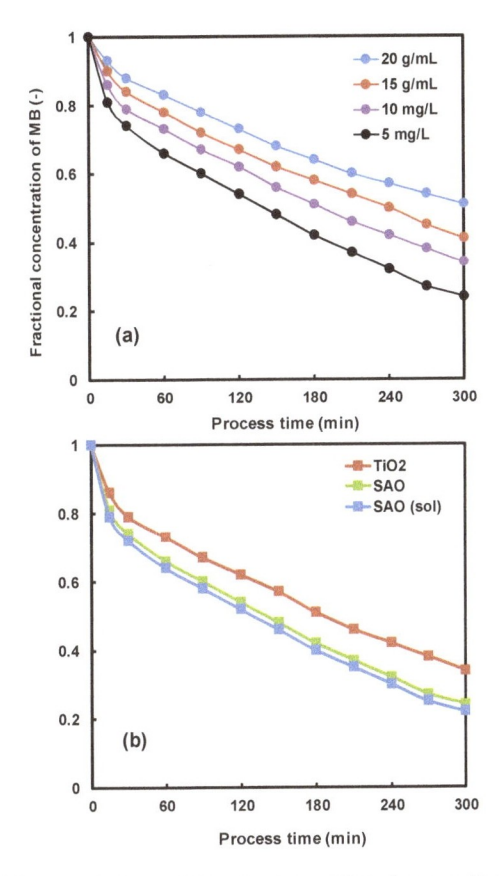

Figure 9. Variation of the concentration of MB by the photocatalytic decomposition on titanium dioxide (TiO_2) with (**a**) various initial concentrations and (**b**) on SAOs at an initial MB concentration of 10 mg/L.

3. Materials and Method

3.1. Preparation of SAOs

The SAOs were prepared by both a sol–gel method and hydrothermal reaction. In both methods, aluminum nitrate nonahydrate ($Al(NO_3)_3 \cdot 9H_2O$, 99%; Duksan, Ansan-City, Korea), strontium nitrate

($Sr(NO_3)_3$, 99%; Duksan, Ansan-City, Korea), dysprosium(III) nitrate pentahydrate ($Dy(NO_3)_3 \cdot 5H_2O$, 99%; Alfa Aesar, Ward Hill, MA, USA), and europium(III) nitrate hexahydrate ($Eu(NO_3)_3 \cdot 6H_2O$, 99%; Alfa Aesar, Ward Hill, MA, USA) were dissolved in distilled water with stirring for 30 min at 90 °C. The chemicals of the reactants were of analytical grade and used as received. The solutions were combined according to the molar ratio of Sr:Al:Eu:Dy = 0.97:2:0.01:0.02.

The chelating reagent solution was prepared using the appropriate amount of aqueous citric acid solution and added dropwise to the above solution. A boric acid solution was then added to the chelating reagent solution. The mixture was concentrated at 80 °C with stirring until it changed to a high viscosity translucent gel. The mixture was then calcined in an electric muffle furnace at 1100 °C for 3 h in a weak reducing atmosphere using a hydrogen-containing gas mixture ($Ar:H_2$ = 95:5).

SAOs were also prepared by a hydrothermal reaction. The mixture of SAO precursors and the chelating agent solution were poured in an autoclave. The autoclave containing the reactant was heated to 130 °C with stirring using a magnetic stirrer. The hydrothermal reaction conditions were maintained for 5 h. After the hydrothermal reaction, the product was dried at 110 °C for 12 h. The product was then calcined at 550 °C for 4 h in a reducing atmosphere. Commercially available TiO_2 (Degussa, P25, Krefeld, Germany) with a particle size and specific area of $\approx$30 nm and $\approx$50 m^2/g, respectively, was also used for photocatalytic decomposition.

3.2. Photocatalytic Decomposition of Methylene Blue

The MB solution (100 mL) was mixed with the photocatalysts as a reactant mixture. The photocatalyst loading was adjusted to 5 mL because of the significant difference in the densities between TiO_2 and SAOs. The reactant mixture was stirred in the dark for 1 h to reach adsorption equilibrium. The reactant mixture was irradiated with UV light with stirring. Samples were taken at regular intervals. They were the centrifuged and the photocatalysts were separated. The concentrations of the samples were analyzed by UV-visible spectrophotometry (Shimadzu UV-2450, Tokyo, Japan). The concentrations of MB were determined from the calibrated absorbance at 665 nm using a spectrophotometer.

The photocatalytic degradation of MB was carried out using a glass reactor-installed UV lamp system. The reactor was kept in the dark to prevent the dispersion of UV light during the photoreaction. The photoreaction temperature was maintained at 25 °C. The UV array consisted of two 10 W UV-A lamps. The UV emission wavelength and light strength was 365 nm and 30 Lx, respectively.

3.3. Characterization of the SAOs

The phase of the SAO particles was determined by XRD (Rigaku Model D/max-II B, Texas, USA). XRD was conducted at 40 kV and 30 mA with a scan speed of 5°/min, scanning angle from 10° to 60°, and a step of 0.02° using Cu Kα radiation. The crystal size and morphology of the SAOs were investigated by SEM (Hitachi S-4700, Tokyo, Japan). Their chemical components were analyzed by EDX (NORAN Z-MAX 300 series, Tokyo, Japan). The N_2 isotherms of the SAOs were investigated using a volumetric adsorption apparatus (Mirae SI, Porosity-X, Gwangju, Korea) at liquid N_2 temperature. The sample was pretreated at 150 °C for 1 h before exposure to nitrogen gas. The surface areas of the samples were calculated using the BET equation [27]. Transmission electron microscopy (TEM, JEOL JEM-2100F, Tokyo, Japan) was performed using a LaB_6 filament and operated at 200 kV.

The composition of the phosphors was analyzed using an EDX micro analyzer (JEOL JSM-840A, Tokyo, Japan) mounted on the microscope. The photoluminescence was evaluated by photoluminescence spectroscopy (PL, Spectrograph 500i, Acton Research Co., Acton, MA, USA) with a 0.2 nm resolution at room temperature. The samples were excited at 226 nm using a He-Cd laser. The UV–vis diffuse reflectance spectra were measured using a UV-vis spectrometer (Shimadzu UV-2450, Tokyo, Japan) in the region, 200–700 nm, with $BaSO_4$ as the reflectance standard. The optical bandgap (E_{gap}) was calculated using the Kubelka–Munk method for indirect electronic transitions.

4. Conclusions

Strontium aluminates co-doped with europium and dysprosium was prepared by a hydrothermal reaction through a sintering process at lower temperatures. The physicochemical properties of the SAOs characterized by SEM-EDX, photoluminescence, and UV-visible DRS were similar to those of the SAOs prepared by the sol–gel method. Although SAOs had been calcined at lower temperatures, their characteristics matched the standard. The photocatalytic activity for the photodecomposition of MB dye was higher than that of the TiO_2 photocatalyst. The SAOs exhibited higher photocatalytic activity than the TiO_2 photocatalyst. The higher photocatalytic activity of SAO was attributed to its higher photosensitivity.

Acknowledgments: This paper was supported by Research Funds of Kwangju Women's University in 2017.

Conflicts of Interest: The author declares no conflict of interest.

References

1. Du, H.; Shan, W.; Wang, L.; Xu, D.; Yin, H.; Chen, Y.; Guo, D. Optimization and complexing agent-assisted synthesis of green $SrAl_2O_4$: Eu^{2+}, Dy^{3+} phosphors through sol-gel process. *J. Lumin.* **2016**, *176*, 272–277. [CrossRef]
2. Amato, C.A.; Giovannetti, R.; Zannotti, M.; Rommozzi, E.; Ferraro, S.; Seghetti, C.; Minicucci, M.; Gunnella, R.; Di Cicco, A. Enhancement of visible-light photoactivity by polypropylene coated plasmonic Au/TiO_2 for dye degradation in water solution. *Appl. Surf. Sci.* **2018**, *441*, 575–587. [CrossRef]
3. Katsumata, T.; Nabae, T.; Sasajima, K.; Matsuzawa, T. Growth and characteristics of long persistent $SrAl_2O_4$- and $CaAl_2O_4$-based phosphor crystals by a floating zone technique. *J. Cryst. Growth* **1988**, *183*, 361–365. [CrossRef]
4. Nakamura, T.; Kaiya, K.; Takahashi, N.; Matsuzawa, T.; Rowlands, C.C.; Beltran-Lopez, V.; Smith, G.M.; Riedi, P.C. High frequency EPR of europium(II)-doped strontium aluminate phosphors. *J. Mater. Chem.* **2000**, *10*, 2566–2569. [CrossRef]
5. Lin, Y.; Tang, Z.; Zhang, Z. Preparation of long-afterglow $Sr_4Al_{14}O_{25}$-based luminescent material and its optical properties. *Mater. Lett.* **2011**, *51*, 14–18. [CrossRef]
6. Singh, T.S.; Mitra, S. Fluorescence behavior of intramolecular charge transfer state in *trans*-ethyl p-(dimethylamino) cinamate. *J. Lumin.* **2007**, *127*, 508–514. [CrossRef]
7. Kubota, S.; Yamane, H.; Shimada, M. A new luminescent material, $Sr_3Al_{10}SiO_{20}$:Tb^{3+}. *Chem. Mater.* **2002**, *14*, 4015–4016. [CrossRef]
8. Lin, C.C.; Xiao, Z.R.; Guo, G.-Y.; Chan, T.-S.; Liu, R.-S. Versatile phosphate phosphors $ABPO_4$ in white light-emitting diodes: Collocated characteristic analysis and theoretical calculations. *J. Am. Chem. Soc.* **2010**, *132*, 3020–3028. [CrossRef] [PubMed]
9. Kiss, B.; Manning, T.D.; Hesp, D.; Didor, C.; Taylor, A.; Pickup, D.M.; Chadwick, A.V.; Allison, H.E.; Dhanak, V.R.; Claridge, J.B.; et al. Nano-structured rodium doped $SrTiO_3$-visible light activated photocatalyst for water decomposition. *Appl. Catal. B Environ.* **2017**, *200*, 547–555. [CrossRef]
10. Liyuan, X.; Qin, X.; Yingliang, L. Preparation and characterization of flower-like $SrAl_2O_4$: Eu^{2+}, Dy^{3+} phosphors by sol-gel process. *J. Rare Earths* **2011**, *29*, 39–43.
11. Chai, Y.S.; Zhang, P.; Zheng, Z.T. Eu^{2+} and Dy^{3+} co-doped $Sr_3Al_2O_6$ red long-afterglow phosphors with new flower-like morphology. *Phys. B Condens. Matter* **2008**, *403*, 4120–4122.
12. Suriyamurthy, N.; Panigrahi, B.S. Effects of non-stoichiometry and substitution on photoluminescence and afterglow luminescence of $Sr_4Al_{14}O_{25}$: Eu^{2+}, Dy^{3+} phosphor. *J. Lumin.* **2008**, *128*, 1809–1814. [CrossRef]
13. He, Z.; Wang, X.; Yen, W.M. Investigation on charging processes and phosphorescent efficiency of $SrAl_2O_4$: Eu, Dy. *J. Lumin.* **2006**, *119–120*, 309–313. [CrossRef]
14. Ayari, M.; Paul-Boncour, V.; Lamloumi, J.; Mathlouthi, H.; Percheron-Guégan, A. Study of the structural, thermodynamic and electrochemical properties of $LaNi_{3.55}Mn_{0.4}Al_{0.3}(Co_{1-x}Fe_x)_{0.75}$ ($0 \leq x \leq 1$) compounds used as negative electrode in Ni-MH batteries. *J. Alloy. Compd.* **2006**, *420*, 251–255. [CrossRef]
15. Peng, T.; Huajun, L.; Yang, H.; Yan, C. Synthesis of $SrAl_2O_4$: Eu, Dy phosphor nanometer powders by sol–gel processes and its optical properties. *Mater. Chem. Phys.* **2004**, *85*, 68–72. [CrossRef]

16. Cordoncillo, E.; Julian-Lopez, B.; Martínez, M.; Sanjuán, M.L.; Escribano, P. New insights in the structure–luminescence relationship of Eu:SrAl$_2$O$_4$. *J. Alloy. Compd.* **2009**, *484*, 693–697. [CrossRef]
17. Rezende, M.V.; Montes, P.J.; Soares, F.M.; Santos, C.; Valerio, M.E. Influence of co-dopant in the europium reduction in SrAl$_2$O$_4$ host. *J. Synchrotron Radiat.* **2014**, *21*, 143–148. [CrossRef] [PubMed]
18. Tao, J.; Pan, H.; Zhai, H.; Wang, J.; Li, L.; Wu, J.; Jiang, W.; Xu, X.; Tang, R. Controls of tricalcium phosphate single-crystal formation from its amorphous precursor by interfacial energy. *Cryst. Growth Des.* **2009**, *9*, 3154–3160. [CrossRef]
19. Zhang, R.; Han, G.; Zhang, L.; Yan, B. Gel combustion synthesis and luminescence properties of nanoparticles of monoclinic SrAl$_2$O$_4$:Eu^{2+}, Dy^{3+}. *Mater. Chem. Phys.* **2009**, *113*, 255–259. [CrossRef]
20. Peng, T.; Yang, H.; Pu, X.; Hu, B.; Jiang, Z.; Yan, C. Combustion synthesis and photoluminescence of SrAl$_2$O$_4$:Eu,Dy phosphor nanoparticles. *Mater. Lett.* **2004**, *58*, 352–356. [CrossRef]
21. Megaw, H.D. *Crystal Structure: A Working Approach*; Saunders, W.B.: London, UK, 1973.
22. Nakamoto, K. *Infrared Spectra of Inorganic and Coordination Compounds*; Wiely: London, UK, 1963.
23. Shan, W.; Wu, L.; Tao, N.; Chen, Y.; Guo, D. Optimization method for green SrAl$_2$O$_4$: Eu^{2+}, Dy^{3+} phosphors synthesized via co-precipitation route assisted by microwave irradiation using orthogonal experimental design. *Ceram. Int.* **2015**, *41*, 15034–15040. [CrossRef]
24. Hong, G.Y.; Jeon, B.S.; Yoo, Y.K.; Yoo, J.S. Photoluminescence characteristics of spherical Y$_2$O$_3$: Eu phosphors by aerosol pyrolysis. *J. Electrochem. Soc.* **2001**, *148*, H161–H166. [CrossRef]
25. Xiao, Q.; Xiao, L.; Liu, Y.; Chen, X.; Li, Y. Synthesis and luminescence properties of needle-like SrAl$_2$O$_4$: Eu, Dy phosphor via a hydrothermal co-precipitation method. *J. Phys. Chem. Solids* **2010**, *71*, 1026–1030. [CrossRef]
26. Kubelka, P.; Munk, F. Ein beitrag zur optik der farbanstriche. *Z. Tech. Phys.* **1931**, *12*, 593–598.
27. Brunauer, S.; Emmett, P.H.; Teller, E. Adsorption of gases in multimolecular layers. *J. Am. Chem. Soc.* **1938**, *60*, 309–319. [CrossRef]

© 2018 by the author. Licensee MDPI, Basel, Switzerland. This article is an open access article distributed under the terms and conditions of the Creative Commons Attribution (CC BY) license (http://creativecommons.org/licenses/by/4.0/).

MDPI
St. Alban-Anlage 66
4052 Basel
Switzerland
Tel. +41 61 683 77 34
Fax +41 61 302 89 18
www.mdpi.com

Catalysts Editorial Office
E-mail: catalysts@mdpi.com
www.mdpi.com/journal/catalysts

Lightning Source UK Ltd.
Milton Keynes UK
UKHW050918181222
413968UK00006B/69

9 783039 282869